Mathematics Projects Handbook

Mathematics Projects Handbook

Third Edition

ADRIEN L. HESS
GLENN D. ALLINGER
LYLE E. ANDERSEN

Montana State University
Bozeman, Montana

NATIONAL COUNCIL OF TEACHERS OF MATHEMATICS

Third Edition

Library of Congress Cataloging-in-Publication Data:

Hess, Adrien L.
Mathematics projects handbook / Adrien L. Hess, Glenn D. Allinger, Lyle E. Andersen. – 3rd ed.
p. cm.
Bibliography: p.
ISBN 0-87353-283-X
1. Mathematics—Study and teaching. I. Allinger, Glenn D. II. Andersen, Lyle E. III. National Council of Teachers of Mathematics. IV. Title.
QA11.H44 1989
510'.712—dc20 89-12330
CIP

Printed in the United States of America

Contents

Introduction

MOST OF US find great satisfaction in exhibiting the results of our study and labor and in comparing our efforts with those of others. Fairs and competitions in all areas of industry and art are familiar features in American life. Such experiences are pleasant and educational for all concerned—the exhibitor learns a great deal in preparing the exhibit, and the viewer learns something new from seeing it.

The project method of teaching was originated more than fifty years ago by John F. Woodhull of Teachers College, Columbia University. It was described as a "normal, indeed ideal teaching process." Teachers of art, home economics, industrial arts, and vocational agriculture have long capitalized on the attraction of a display of articles or objects produced by students to make parents and friends aware of the work being done in their classrooms. Eventually, science teachers also saw the value of student displays. Such exhibits were usually projects done during regular class time or under a directed program.

Mathematics teachers have also long valued the use of mathematical models and projects to arouse interest in mathematics. These ideas have gained momentum because of the growth of science fairs and the emphasis on mathematics stimulated by military and space-exploration rivalries and by advances in technology. The proportion of mathematics projects at science fairs has been increasing, and school mathematics fairs for students have been conducted at all levels, beginning with the first grade. Today, the uses of technology (calculators, computers, videodisks, etc.) in the mathematics classroom offer wonderful opportunities for student projects.

This handbook is designed as a guide for teachers and students in choosing and developing mathematics projects, from simple demonstrations of mathematical problems or principles that the teacher has assigned as classroom learning experiences to complex, sophisticated exhibits intended for entrance in fairs and competitions.

The third edition updates the references while still including classical texts and journal articles. New topics such as fractals are included, and greater emphasis has been placed on topics of current interest, such as discrete mathematics, probability, and statistics. In addition, reproducible student guides for mathematics projects and fairs have been included.

1

Developing a Mathematics Project

A MATHEMATICS PROJECT consists of all the effort expended in solving a problem, exploring an idea, or applying a mathematical principle—that is, the initial planning, the study, the exhibit, and the written report. It should develop a new mathematical concept or theorem, show the relation of a mathematical idea or principle to some other branch of mathematics or science, or demonstrate the application of a mathematical idea or principle. The exhibit uses drawings, graphs, models, pictures, words, and so on, to briefly tell the viewer the student's idea of the mathematical concept or principle, to show its use in explaining some biological or physical phenomenon, or to answer a question posed by the author of the project. The exhibit can be a collection, an experiment, or a working model; it may be a replication of an earlier experiment or an entirely original work. Thus the *project* is the whole process, whereas the *exhibit* is the tangible attempt to show and explain the project to anyone who is interested.

Choosing a Project

The best projects grow out of students' interests; therefore, they should be chosen by the students themselves. The projects selected should express the students' broadest knowledge and highest skills; they should be enjoyable to do. Teachers' suggestions, conferences, workshops, demonstrations, lists of previous projects, and so on, are helpful in stimulating interest. Chapter 2 gives some titles of previous projects and suggestions for other projects.

Research Methods

As soon as a specific idea or problem has been chosen, it should be thought through carefully. The successive steps should be enumerated and possible pitfalls noted. A written record should be kept at every stage of the project.

When the preliminary plans are made, the student should begin a comprehensive reading program to broaden his or her understanding of the possibilities and limitations of the proposed project. Extensive use should be made of books and periodicals from all available sources.

Since accurate notes should be kept of all readings, observations, suggestions, and speculations, paper and pencil should be handy at all times. The notes should include a complete record of sources. For example:

Authors ↓ — **Title** ↓

Hess, Adrien L., and Carl Diekhans. "The Number of Segments Needed to Extend a Cube to *N* Dimensions." *Mathematics Magazine* 43 (September 1970): 187–92.

↑ **Pages** — ↑ **Journal** — ↑ **Volume** — ↑ **Date**

Hess, Adrien L. *Four-Dimensional Geometry—an Introduction.* Reston, Va.: NCTM, 1977.

↑ **Publisher** — ↑ **Copyright date** — ↑ **Book title** — ↑ **Place of publication**

Ideas should be jotted down even if they seem irrelevant at the time. They can be better evaluated after they have "simmered" for a while.

As the reading and study progress, plans can be discussed with other people; discussing an idea with someone else often gives a new and clearer perspective. Parents, teachers, professional mathematicians, and other students can often make comments, criticisms, or suggestions that are very constructive.

In all research, experimentation, and study, the student should keep in mind the last step in the preparation of the exhibit—the completion of the written report. If all ideas (original or not), guesses (good or bad), measurements, sketches, and reading are recorded in a notebook as the project develops, the final report will be much easier to write. It should include the following:

1. Title—accurate and not too long
2. Introduction—a brief statement of the idea, the principle or problem, and the reason for the student's interest in it (Background information should be given, including relevant work done by others. If little or no work has been done previously, this fact should also be noted.)
3. Discussion of the problem, idea, or principle being investigated
4. Details of materials, equipment, methods, and steps used in solving the problem
5. Summary of results and conclusions
6. New questions, possible applications, future plans
7. Appendix—graphs, tables, photographs, drawings, computer programs
8. Bibliography and acknowledgments

After the exhibit and the written report are completed, the student should consider what questions the judges or interested spectators might ask. He or she should be able to explain the project both to the judges, who may be experts in the field, and also be able to explain it to a layperson, who may be interested but know nothing about the subject.

Construction Materials and Methods

Ingenuity in selecting and using materials is an important factor in the success of an exhibit. For a project that requires a computer, a great amount of the material is already assembled. However, most mathematics projects use a wide variety of materials, such as balsa wood, cardboard, cork, drawing paper, Masonite, metal, papier-mâché, pegboard, plaster of paris, and plastic. Styrofoam can easily be cut with a razor blade or a sharp knife and fastened with pins and airplane glue. Metal and wire must be cut with a hacksaw, tin snips, or pliers; such material makes a sturdy exhibit when soldered. Papier-mâché and plaster of paris can be used for very rigid models. String works well for conic sections and quadric surfaces, and a light source can be used with string models to produce interesting projections. Drawing and lettering can be done with stencils, ink, paint, or felt-tip markers. Manufactured letters can be used for important lettering, such as the title of the project. Another alternative is to do lettering on a computer.

The exhibit must be designed with several things in mind. From the standpoint of its maker, the exhibit should be—

- functional: it should tell the story of the study and research., and any displayed material should be in harmony with this objective;
- practical: the number of components, the time needed to set them up, and any limits on the contents must be considered;
- portable: the exhibit should be compact, durable, and easy to transport.

Responsibilities and Guidance

The Responsibility of the Student

Preparing a mathematics project can be an interesting and worthwhile experience. Students will get experience in using resources to find information, in doing independent work, in organizing their presentation, and in communicating ideas orally, visually, and in writing. They will broaden their backgrounds in mathematics and explore mathematical topics they never knew existed. In setting up exhibits, they will experience the satisfaction of demonstrating what they have accomplished. Students will grow in self-confidence as they share their experiences with others. They will satisfy their curiosity and their desire to be creative. They will develop originality, craftsmanship, and new mathematical understanding. Preparing the project can help them decide if they want to make mathematics their career. At the same time it will dramatize the nature of mathematics to students, teachers, and citizens of the community.

Exhibits are usually judged on the following criteria: originality, completeness, clarity, interest value, craftsmanship and mathematical thought. In organizing a successful exhibit, students will find the following suggestions helpful:

1. *Select a topic that has interest potential.* The topics listed in chapter 2 are suggestive of the many possibilities.

2. *Find as much information about the topic as possible.* Check journals, books and cumulative indexes, such as those listed in chapter 4. Computer searches can be undertaken at local libraries to access books and journals anywhere in the country.

3. *Prepare and organize your material into a concise, interesting report.* Include color drawings, pictures, applications, and examples that will get the reader's attention and add meaning to your exhibit. Build models, mock-ups, or other devices that create interest and explain the topics.

4. *Build an exhibit that will tell the story of your topic.* Use models, applications, or charts that lend variety. If possible, prepare materials that viewers can manipulate. Give your exhibit a catchy, descriptive title. Label everything with brief captions or legends so that viewers will understand the principles involved. Make the display simple but also attractive and dramatic. Use color for emphasis. Write captions in a unique way with such materials as with rope, pipe cleaners, plastic tubing, or yarn. Show craftsmanship, creativity, and diligence in arranging the exhibit. Have available for distribution a brief summary of the basic ideas, plans, and references for your topic.

5. *Be able to demonstrate the topics of your exhibit.* Speak clearly and correctly. Be well informed so that you can answer questions. Be well dressed, courteous, and congenial to the viewers as well as the judges.

The Role of the Mathematics Teacher

One of the most important functions of the mathematics teacher is to provide enthusiasm and inspiration so that students will *want* to do a project. The teacher should have available a wide selection of ideas and suggestions that will be suitable for mathematics projects: newspaper clippings on mathematics winners; lists of the project titles of previous winners; descriptions and reports of previous mathematics projects; and slides, videotapes, and photographs of winning projects. With the help of the school librarian, the teacher can provide a well-rounded selection of books, pamphlets, and periodicals on mathematics.

The teacher should be ready to help the student choose a project and should review the criteria for it with the student before the work actually begins. The project should be one in which the student has an interest, and at the same time it should be within his or her mathematical ability.

The teacher should be able to offer guidance throughout the preparation of the exhibit and should maintain a cheerful and optimistic attitude at all times, especially when the student is not making the progress he or she desires.

A "project progress report" should be requested from each participant. Included in this report is a time line for completing the project. (See Appendix A for a progress project report form that can be duplicated.)

Finally, the teacher should check the finished exhibit and project report for the correct use of mathematical principles and terms and for accuracy in grammar and spelling.

The Role of the Parents

Studies of winners in science fairs and talent searches have shown that children's interest in mathematics and science is strongly influenced by their parents. If parents provide a family atmosphere that includes the enjoyment of learning and stimulating conversation, the child may develop such important habits and traits as independence, intellectual curiosity, perseverance, responsibility, and creativity. A project is the student's work, but the parents can be supportive by showing genuine interest. They should be generous with both encouragement and criticism as the occasion demands.

Parents should provide the proper tools, materials, and an adequate place to work, and they should help with arrangements to transport the exhibit to a location for display.

The Role of the Judges

The men and women who take time from their busy schedules to judge projects do so because they are interested in youngsters and in mathematics. Ideally, the judging should be divided into three steps.

1. Judges should tentatively evaluate each exhibit independently without the exhibitors or the public being present. During this examination they can make notes, including questions to ask, points to discuss, and improvements to suggest to the exhibitor.

2. After this preliminary examination, the exhibitor should be interviewed. This interview is one phase of the learning process and should not be omitted. Judges should be ready and willing to give suggestions and ideas to the exhibitor and ascertain what he or she has really learned.

3. In the final judging, the evaluation reports should be examined, and the individual ratings can be discussed among the judges, if they wish.

Evaluation Criteria

It is very helpful to have certain criteria and standards to use as a guide in evaluating projects. The *Sponsors Handbook, 1963–64 Supplement,* prepared by Science Service (1719 N Street, NW, Washington, DC 20036; tel. (202)785-2255) has not been updated and is still the guide used to judge all projects entered in the Westinghouse Science Talent Search. It is a good idea to review with the participants a few times during the preparation period the following criteria and standards for judging (e.g., see p. 25 in the *Sponsors Handbook, 1963–64 Supplement*):

1. Creative Ability (30 points)

How much of the work appears to show originality of approach or handling? Judge that which appears to be original regardless of the expense of purchased or borrowed equipment. Give weight to ingenious use of materials, if such exists. Consider collections creative if they seem to serve a purpose.

This first criterion weighs heavily in an evaluation and can cause more worry than any of the others. Everyone is creative to some extent, but creativity can be developed through hard work. Students can start by reading a book or two on creative thinking. Some of the different kinds of creativity are (*a*) discovering relationships between facts that seem unrelated, (*b*) observing new unifying factors in groups of facts or observations, (*c*) finding useful applications in one field and transferring them to another field, and (*d*) recognizing the importance of something stumbled onto by accident while looking for something else. Flashes of insight usually come only after considerable study and hard work, and often after numerous failures.

2. Scientific Thought (30 points)

Does the exhibit disclose organized procedures? Is there a planned system, classification, accurate observation, controlled experiment? Does the exhibit show a verification of laws or cause and effect or present by models or other methods a better understanding of scientific facts or theories? Give weight to the probable amount of real study and effort that is represented in the exhibit. Guard against discounting for what might have been added, included, or improved.

The quality of scientific thought, as defined for science projects, may need to be redefined for mathematics projects. Much of scientific thought involves examining a question relating to nature. It is possible that this same approach can be used in some mathematics projects, but in others it will need to be modified to obtain a result that is mathematically sound.

3. Thoroughness (10 points)

Score here how completely the story is told. It is not essential that a step-by-step elucidation of construction details be given in the working model.

4. Skill (10 points)

Is the workmanship good? Under normal conditions is the exhibit likely to demand repairs? In collections, how skilled is the handling, preparation, and mounting?

5. Clarity (10 points)

In your opinion will the average person understand what is being displayed? Are guide marks, labels, and descriptions neatly yet briefly presented? Is there a sensible progression of information that directs the spectator's attention through the exhibit?

6. Dramatic Value (10 points)

Is this exhibit more attractive than others in the field? Do not be influenced by such "cute" things as lights, buttons, switches, cranks, or other gadgets that contribute nothing to the exhibit.

The dramatic value may count for much more than the number of points allocated to it as a criterion. Motion, color, and sound are ways of adding dramatic value. All labeling and lettering should be carefully, correctly, and neatly done. The title should be displayed briefly but well.

After the Exhibit

A student who has one or more years remaining in junior high or high school may want to continue working on the exhibit the following year. It often takes more than one year to produce an outstanding exhibit. Senior students may wish to donate their exhibits to the mathematics department.

Service clubs and parent-teacher groups are usually willing to have a student discuss his or her exhibit at one of their meetings: the student not only gains experience in discussing the project but informs the public of school activities as well.

If the written report is of sufficient interest, it can be submitted by the participant and teacher for publication in professional journals such as the *Mathematics Teacher*. See chapter 4 for a list of other periodicals and for additional information about the development and exhibition of mathematics projects.

2

Looking for Ideas

THERE IS NO one source of ideas for projects; an idea can come in various ways and from many sources. Conferences and conversations with teachers, mathematicians, scientists, and other students are sources of suggestions. Ideas can also be obtained from listings of previous mathematics projects, from reports of fair winners in talent searches, from articles in magazines, and from newspaper stories concerning mathematics. Mathematical and scientific books and periodicals are a fertile source of ideas for projects. An idea can begin with a statement in an article, a question, or a reference that causes the student to wonder why or why not and motivates him or her to investigate the problem further. Obviously, anyone looking for an idea should be an avid and regular reader of a wide range of mathematical and scientific periodicals.

Titles of Projects

The following classified list of project titles, all of which have been entered in a science fair or in a science talent search, can be used by students to give them some direction in their search for potential projects. The same title might suggest entirely different projects to different people.

HIGH SCHOOL

Algebra, Game Theory, and Theory of Numbers

- Algebra of Sets
- Divisibility
- Polyominoes
- New Relations for Fibonacci-like Series
- Properties of Fibonacci-like Sequences
- Development of Class-Two Perfect Numbers
- Consecutive Prime Powers
- Unitary Superperfect Numbers
- An Investigation of the Twin Primes Problem
- Natural Logarithms of Complex Numbers
- Convergence of Continued Fractions
- Special Matrix Exponential Forms
- An Examination of the $XN + 1$ Problem
- Derivation of the General Form of Divisibility Functions for Types of Moduli
- Densities of Non-Deficient Numbers in Several Integer Sequences

- Proof of Some Properties of Fibonacci Numbers
- Demonstration of the Inadequacy of Induction in Mathematical Proofs
- Commutative Ring and the Set of Multiplicative Structures in an Additive Abelian Group
- The Use of Game Theory to Determine Courses of Action
- Accounting for the Relationship of the Fibonacci Sequence to Nature's Mathematical Spirals
- An Investigation of Pythagorean *N*-tuples and Similar Diophantine Equations
- The Covariant Chain Rule of Fiber Bundles
- Combinatorics—the Art of Counting
- Rotation Groups
- New Approach to Quadratic Equations
- General Solution of a Linear Diophantine Equation of Three Variables
- Matrix Inversion by Means of the Cayley-Hamilton Theorem
- Consecutive Integral Triangles
- The Number of Ways to Represent the Reciprocal of an Integer as the Sum of the Reciprocal of Two Integers
- Factorable Sets of Integers of the Form $ak + b$
- Matrix Representation of the Chromosome Theory
- A Study of Prime Numbers
- Rational Polynomial Functions
- Electronic Graphical Representation of Any Second-Degree Equation
- Solution of Certain Diophantine Problems by Functional Methods
- Convergence of Continued Fractions
- The Binary Group
- Proof of Some Properties of Fibonacci Numbers
- Factors Determining the Nature of Multiplication Tables for Other Bases
- Permutations
- Experimental Treatment of Square Root as Infinite Geometric Progressions
- Hypercomplex Numbers
- A Mathematical Approach to Knot Typing
- Linkages and the Application of Their Properties to Solving Equations
- The Pythagorean Triples
- Properties of Pythagorean Triples Where Legs Differ by One
- The Use of Game Theory to Determine Courses of Action
- A Graphical Solution of 3 ¥ *m* Game Matrices
- Group Theory of the Equilateral Triangle
- Spherical Functions of Ordered Pairs
- Constructing a Mathematical Ring
- Multinomial Expansion
- Proof of a Mathematical Theory in the Generation of Langford's Numbers
- The Function $x^y = y^x$
- Generation of Magic Squares and Magic Cubes
- A Study of Fermat's Equation
- A Topic in Diophantine Analysis: Cole Numbers
- A New Graphic Study of Magic Squares of the Fourth Order
- Pauli Matrices as an Example of a Group
- Homeomorphisms
- Bicomplex Numbers
- Theory of Braids
- Think You Can Beat Game Theory?
- Multiplication Table for a Noncommutative Group
- Algebraic Solution to Electrical Circuits
- Boolean Algebra
- Finite Sums of Polynomials
- A New Sieve for Finding Prime Numbers
- Analysis of Algebraic Congruences in One Unknown
- Investigation of Odd Perfect Numbers

- On the Number of *N*th Roots in Finite Abelian Groups
- Artificial Intelligence
- Symmetric Functions of Roots
- Analysis of Symmetric Groups of Degree Four and Their Relation to Eigen Values
- Graphing $f(x) = x^{1/2}$
- Use of Logarithmic Scale in Graphs
- Exponential Functions
- Graphical Representation of Complex Roots of Quadratic and Cubic Equations
- Flip Side of Fibonacci
- Fibonacci-type Sequences with Continued Fractions
- Mathematics and Switching Circuits
- "Proof" of Goldbach's Conjecture via Probability
- Symmetry in Equations
- Modulo Arithmetic
- An Algebraic Approach to the Fourth Dimension
- Investigation of Moduli Systems and Pascal's Triangle
- Elementary Random Number Generators
- An Investigation of Perfect Numbers Both Even and Odd
- Inequalities Related to Fibonacci Numbers
- Fifth-Dimension Analog of a Binomial Cube
- Algebraic Balancing of Chemical Equations
- Development of a General Semigroup

Analysis

- Transfinite Mathematics
- Convex Smooth Curves
- A Nondifferentiable Function
- Minimal Surface Area Experiments with Soap Films
- Hyperbolic Functions and Related Infinite Series in the Complex Domain
- A Development of a Power Series for the Sum of the *r*th Term of the First *N* Numbers
- Application of Calculus to the Solution of Game Matrices
- A New Method of Summing Certain Infinite Series
- Arc Length
- Elliptic Integrals
- The Brachistochrone-Tautochrone Problem
- The Calculus of Variations
- Development of a Term Difference Method for Function of Finite Sets
- Arranging Fractions between 0 and 1 in Ascending Order of Numerical Value
- $F(x) = F(1 - x)$ and Other Mathematical Investigations
- Evaluation of Real Integrals by Means of Complex Residues
- Empirical Variation on Newton's Method
- Diagonal Sums of General Slope in a Right Pascal Triangle and Their Relation to Fibonacci-type Sequences
- Ordering the Set of Complex Numbers
- Calculus of Finite and Infinitesimal Ratios
- The Concept of Number Applied to Infinite Sets
- Application of Cybernetics to the Solution of Differential Equations
- *N*-Point Tensor Calculus
- Curvature of Droplets
- Comparison of Set Theory with Elements and Compounds
- A Multiplicative Calculus
- A Discussion of Large Numbers with Special Application to Infinities of Varying Size
- Logical Analysis of Infinity and Infinite Set Theory

Arithmetic

- Elementary Binary Arithmetic
- Mathematics to Base n ($n \neq 10$)
- Binary Logarithms
- Binary Counter
- From Sticks to Numerals
- Roman Numerals and the Abacus
- The Imaginary Binary System
- Comparison of Number Systems
- World's Oldest Adding Machine and How It Works
- Magic Number Guessers—Binary Notation and Intersection of Sets
- General Form of the Arithmetic Processes and Its Extension to the Process after Passing to a Power
- Extensions of Arithmetic Operations
- Numerical Taxonomy
- Napier's Rods
- The Duodecimal System
- "Trigesimal" Number System
- The Algorithm of Euclid
- Numerology
- Informal Study of Digital Numbers
- An Arithmetical Method to Find the nth Root of a Number
- A Number System to an Irrational Base
- Investigations in the Theory of Decimal Expansions
- Computation in Systems Other Than Decimals
- An Arithmetical Investigation of Squares and Their Relationships
- Analysis and Application of Binary Numeration
- Digital Analog to the Ancient Abacus

Computers, Logic, and Numerical Analysis

- Set Theory with Application to Symbolic Logic
- Basic Computer Components
- Binary Computer to Find Square Root
- "Thinking Machines" and the Thought behind Them
- Derivation of Rules for Programming the Game of Tic-Tac-Toe
- Symbolic Compiler for Arithmetic and Logical Programs
- Compilation and Study of First 2562 Decimal Places of e
- Digital Classification of Words for Decoding
- The Braille-Scuber—an Original Use of a Digital Computer
- Approximation to Euler's Constant by Probability Theory and Random Numbers Applied to High-Speed Electronic Computer
- A Decimal-to-Binary Converter and Binary Adder
- Design and Construction of Two Computers: The Quadralicator and the Geometron
- An Electronic Analog Computer for the Solution of Cubic Equations by Cardan's Formula
- Mathematical Investigation of Substitution Cipher
- Ciphers, Codes, and the Way They Are Broken
- A Simple Logic System
- A Binary Relay Computer and Simple Transistor Circuits
- Calculating Machines of the Future
- Design, Development, and Construction of a Small Binary Digital Computer
- Electronic Nim Partner
- Stonehenge—a Neolithical Calculator
- Design and Construction of Several Binary Digital Computers
- Digital Quadratics Computer
- Number Systems and Computers
- The Semiautomatic Binary Digital Computer
- An Analog Computer to Solve Combined Charles's and Boyle's Law

- Formation of Digital Root Series
- Cybernetic Computers
- An Electrical Computer Capable of Proving Thousands of Geometric Theorems
- An Electronic Syllable Analyzer
- An Application of a Digital Computer to the Solution of Rectangular Games
- Electronic Analog Computer

Geometry

- Finite Projective Geometry and Abelian Groups
- Quasi-Brocard Geometry
- Excursions into Experimental Geometry
- Algorithmic Methods for Generating Reverse Theorems
- Locus by Light
- Plane Projections of the Earth
- The Hyperbolic Paraboloid
- Which Figure Has the Greatest Area?
- Biangular Systems of Coordinates
- A Finite Geometry
- The Superbola $Y = X^X$
- Mascheroni's Constructions
- Finding the Area of an Ellipse
- Three-Dimensional Dominoes
- The Finite Solid Geometry System
- Duality in Points and Lines
- Locus in Space
- Geometry of a Catenary
- Bipolar Geometry
- The Shape of Place Intersection
- The Great Icosahedron
- The Triangle Theorem of Desargues
- Fractals
- Investigations into Areas of Fractal Patterns
- Pass Filtering on Fractal Dimension
- Fractal Nature of a Chaotic Attractor
- Viscous Fingers as Fractals
- A Continuous Conic Section Generator
- The Hyperboloid
- Euclidean Constructions with Arbitrary-sized Instruments
- More about Perimeters
- Place Intersections in *N* Dimensions
- Geometric Constructions with a Tool Other Than the Compass and Straightedge
- Pascal's Pyramid
- Conic Sections and Allied Surfaces
- The Five Polyhedrons
- Golden Section
- Investigating the Nine-Point Circle in Three-Space
- An Extension of the Theorem of Pythagoras
- A Geometry of the Triangle of Progressions
- Nine Postulates for Euclidean Geometry
- Primitive Geometry Taken from the Indians
- The Basic Theory of Perspectives
- Limitations on Euclidean Geometry
- Geometry of Bubbles and Liquid Film
- Geometric Designs
- String Designs
- Chameleonic Cubes
- Space Mathematics
- Geometric Models
- Multidimensional Geometry
- The Integral Right Triangle
- Theory of Rotations of Coordinate Axes in *N*-Space
- Angular Coordinates
- Graphs of Conic Sections for Complex Values of the Variable
- Solving the 60° and 120° Triangle
- Mathematics of Crystals
- Quadric Surfaces
- Quadratic Equations with Two Variables Derived from Conic Sections
- A Comparison of Three Geometric Transformations

- How Eratosthenes Measured the Circumference of the Earth
- Euclid's and Lobachevski's Assumptions on Parallels
- Determinatal Formula for the Area of the Two-Dimensional Polygonal Figures
- Folding Polyhedrons
- Analytic Geometry of Oblique Coordinates with Special Attention to the Plotting of Inequality Relationships
- Finding the Equation for the Inversions of a Curve Centered at the Origin
- Geometric Foundations of the Theory of Relativity
- Analytic Geometry in *N* Dimension
- Mathematical Analysis of the Locus of Points on a Rotating Reuleaux Polygon
- Investigation of the Trochoid Family of Curves
- Cylindrical Projections and Their Applications
- Conjugate Coordinates in the Study of the Cardioid
- Plotting a Cubic Equation in Three Dimensions
- Solving a System of Three Quadratic Equations by Means of a Three-Dimensional Graph
- Determination of Equations of Complex Curves Surfaces in Space
- Aristotelian Space—a Research into the Logico-Prospective Properties of Pascal's Table
- Geometric Dissections—Tangrams
- Brocard Points in Aviation
- Centroids of Plane Figures
- Gergonne's Problem
- Projected Geometric Progressions
- An Extension of the Classical Problem of Apollonius into Three Dimensions
- The First Twelve-Point Sphere of an Orthogonal Tetrahedron
- Vector Applications to Ratios in Triangles
- An Investigation of Periodic Linear Fractional Transformation
- Geometric Implications of an Algebraic Equation
- Developing Three-Dimensional Space-Time Geometry
- Multidimensional Evolution of Certain Polygons
- An Extension of Euler's Polyhedral Formula to *N* Dimensions
- Pyramids
- Hyperbolic Plane Geometry
- Furthering Pappus's Extension
- Geometry on a Cylindrical Surface
- Production of Geometric Forms with Mass
- Studying the Twist and Turn of a Third-Dimensional Curve
- Finite Coordinate Systems on Surface of a Sphere
- Geometric (and Relativity) Concepts of Four Dimensions
- Fundamentals of Vector Analysis
- Basic Postulates and Theorems in Synthetic Euclidean Fourth-Dimensional Geometry
- Geometry of *N* Dimensions with Emphasis on Topologies
- Graphing in Four Dimensions
- Fourth Dimensional Tetra-Quadric Surfaces and Their Application
- Fourth-Dimension Space-Time Continuum
- *N*-Dimensional Conic Sections
- The Quartic Curves
- Shells and the Geometric Spiral
- Theorems in Curved Space
- A Proof of Cavalieri's Theorem
- Optics of the Ellipse with Emphasis on Parallel Incident Rays
- Bubble Curves and the Roulettes of Conic Sections

- The Realm of the Spherical Triangles
- Measurement of the Earth and Distance to the Moon by Simple Geometry
- Pascal's Triangle in Three Dimensions
- Pappus's Extension of the Pythagorean Theorem
- Flexagons
- Geometric Interpretation of Super-Perfect Numbers
- Geometry of Parangular Complex Inverse, and Complex Inverse Polygons
- Curves of Constant Width

Statistics and Probability

- How to Lie with Statistics
- The Number *e* by Spinning a Needle
- Mathematical Aspects of Population Growth
- Mathematical Probability and Mendel's Law
- A Statistical Analysis of Background Radiation in a Selected City
- Statistical Study of Finger Length Variations in Adolescent Hands
- The Error Curve in Modern Science and Mathematics
- A Statistical Analysis of Fossil Species
- Driving Fatalities
- Triangular Proof of the Law of Probability
- The Correlation between Mathematics and Musical Ability
- Statistical Analysis of Temperature Variations with Distance
- Probability in Genetic Domination
- A Vector Approach to Statistics
- Mathematics and Games of Chance
- Mathematical Analysis of Batting Performance in the Game of Baseball
- Polynomials of Best Fit by the Method of Least Squares
- A Statistical Study of Auto Engine Repairs
- A Statistical Study of Some of the Factors That Influence People in Buying
- Determination of π by Probability
- Geometric Figures in Probability
- The Tetrachoric Correlation

Trigonometry

- Graphing Trigonometric Functions on Rectangular and Polar Coordinates
- Trigonometric Functions from the Unit Circle
- Minkowskian Trigonometry
- Trigonometric Curves and the Unit Circle
- Generalized Trigonometric Functions and Spiral Trigonometry

Topology

- Combinatorial Topology
- One-Sided Surfaces
- A Discussion of the Four-Color Problem
- Topology, One-sided Surfaces, and the Königsberg Problem
- Relation of Continuity between Regions of a Map Applied to the Proof of the Four-Color Problem
- What Is Topology?
- A Topological Study of Paradromic Curves
- Construction of a Klein Bottle
- Proof of Euler's Formula and Its Use in Proving the Five-Color Problem
- Variation of Möbius Strip and Other Problems of Topology
- Topological Analysis by Means of Dual Maps
- Solution of a Famous Eight-Color Problem and Its Generalization
- "Rubber Geometry"—Some Properties of Topological Surfaces
- Topology of Knots

Miscellaneous

- Mosaics by Reflections

- Mathematics in Music and Sound
- Chinese Rings
- Function before Fashion—Mathematical Designs
- A Mathematical Theory of Relativity
- Mathematical Principles of Particle Acceleration
- Models and Mathematics of the Vanguard Rocket
- Analysis of Hurricane Paths
- Mathematical Models of Yeast Growth
- Attempted Proof of the Four-Color Problem Using Mathematical Induction
- Relationship between Mathematical Formula and Marine Mollusk Shell Growth
- Mathematics of the Suspension Bridge
- Harmonic Motion
- Recursive Patterns in Leibnitz Harmonic Triangles
- The Magic of Mathematics
- Mathematical Study of Lissajou's Figures
- Automatic Computation of Celestial Orbits and Trajectories
- Evaluation of DeSaus's Double Decimal System
- The Properties of a String
- A Hydrostatic Equation Solver
- Mosaics: Math Invades Art
- Hyperboloid High-Rise Structure
- Original Methods of Mathematical Notation

JUNIOR HIGH SCHOOL

- World-wide Foundations of Mathematics
- The Abacus versus the Calculator
- Roots by Rates
- Proof by Nines
- Pascal's Triangle
- Geometric Mobile
- A Fourth Dimension
- Geometric Solids
- Volume of Solids
- From Squares to Circles
- History of the Calendar
- Finding Inaccessible Distances
- Three-Dimensional Graphings
- Geniac Experiment
- Optical Illusions
- Geometry by Paper Folding
- Indirect Measurement
- Boolean Algebra
- Analog Computer
- A Comparison of the Septimal and Decimal Systems
- The Abacus
- Decision Maker
- Symmetry in Nature
- Binary System
- Add Instead of Multiply
- Number Systems
- Napier's Rods
- Pi
- Logo—a Fun Computer Language

Ideas for Reports and Things to Construct

Many students may not have the desire or the time to do a project like those discussed earlier. They might, however, find it interesting to do a report, either oral or written, or to construct some mathematical curve, device, or instrument. Such reports can be a part of classwork or of a program for a mathematics club and might encourage more student interest in mathematics.

The biographies of famous mathematicians, both men and women, can prompt interesting and informative reports. Some of the men who made outstanding contributions to mathematics before the twentieth century are Abel, Archimedes, Cantor, Desargues, Descartes, Eratosthenes, Euler, Fermat, Galois, Gauss, Leibniz, Lobachevski, Newton, Pythagoras, Riemann, Saccheri, and Thales. Modern mathematicians include Birkhoff, Blackwell, Chern, Conway, Coxeter, Diaconis, Erdos, Gardner, Graham, Halmos, Hilton, Kemeny, Kline, Knuth, Mandelbrot, Pollak, Pólya, Robbins, Smullyan, Tucker, Ulam, and Whitehead. These names can be found in encyclopedias or in histories of mathematics. See, for example, Burton (16); Eves (30, 31); Christianson (17); Lowe (77); and Albers (3)).

Women have made significant contributions to the development of mathematics from ancient times. Historically, however, little attention was given to their work. A noticeable change in society's recognition of, and appreciation for, women's contributions to mathematics has occurred only within the last three or four decades. Among the outstanding women in mathematics are Agnesi, du Chatelet, Germain, Herschel, Hypatia, Kovaleski, Noether, Rees, Reid, Rudin, Sommerville, Taussky-Todd, and Young. See Albers (3); Burton (16); Eves (30;31); Fox (33, 34); Koblitz (68); Osen (94); Perl (97); Stein (128).

A number of people have made exceptional contributions to mathematics even though they were not professional mathematicians (i.e., their main interest area was outside mathematics.) Some of these are Dürer, Napier, Pascal, Buffon, and Bolzano. These names can be found in various references, including Eves (30;31).

Statistics is currently one of the fastest growing branches of the mathematical sciences, and it is a relative newcomer to the K–12 school curriculum. The ideas of probability and statistics come from the earliest mathematicians. David (24) discusses their development through the Newtonian era. Names he includes are Bernoulli, Cardano, DeMoivre, Fermat, Galileo, Huygens, Pepys and Wallis. Box (13) details the life of R. A. Fisher, often called the father of modern statistics. Some other twentieth century female and male statisticians are Bailar, Box, Cochran, Cox, Deming, Hunter, Martin, Mosteller, Norwood, Scheffe, Snedecor, and Tukey. Other sources are Pearson (96) and Walker (137).

The construction of geometric figures is interesting and motivating to many students. Straightedge-and-compass constructions are easy to do and produce very pleasing designs. If a greater challenge is needed, restrictions can be placed on the type of tools used or the method of using them. Some of the easiest constructions are regular polygons, triangles, and circles that contain hexagonal figures. The construction of regular polygons with strips of paper is intriguing. More complex curves include cardioids, catenaries, cissoids, conchoids, and cycloids. The problem of Apollonius, which involves circles, will also appeal to some students. Numerous points and lines encountered in the study of geometry are of interest, such as Euler lines, Simson lines, Brocard points, and Gergonne points. Some of these can be studied in connection with the nine-point circle. The construction of three-dimensional models of polyhedrons can be interesting. For information on constructions, see Courant (21); Cundy (23); Hilton (53); Hobson (55); Kenney (65); Kostovskii (71); Lyng (79); NCTM (151); Moise (86); Posamentier (101); Seymour (114, 115, 116); Smart (118); Wenninger (139, 140); and Yates (147). In addition, computer programs that use graphics, such as the Geometric Supposers by Schwartz (111, 112) and Logo (9), with its turtle geometry, are also effective media through which students can explore and discover geometry concepts.

Many aspects of mathematics have applications in such areas as art, music, news, cartoons, nature, or photography. Other applications deal with how mathematics is used in a particular occupation or how it has affected physical theories, political thought, social science, or ancient cultures. Since these topics are so diverse, a student must read widely from books, newspapers, and magazines. Cundy (23) is a good place to begin. Teachers will want to consult NCTM's 1979 Yearbook, *Applications in School Mathematics,* as well as Kastner's *Applications of Secondary School Mathematics* and MAA and NCTM's *A Sourcebook of Applications of School Mathematics,* all of which are listed in the "Resources" section of this book. See Bezuszka (8); Haak (173); Hoffman (177); Kline (67); Kluepfel (180); Laffan (181); Lund (182); Phillips (99); Schroeder (110); Selkirk (113); Slawsky (190) and Townsend (133).

Students will find that certain constants—their history, their relationships and their occurrence—are very interesting. Some of the more common constants are 0, 1, i, e, π, and ϕ. The first five of these constants all occur in the equation $e^{i\pi} + 1 = 0$, where $i = \sqrt{-1}$). In the Hindu-Arabic numeration system, 0 and 1 are the first two digits; 0 is the identity element for addition, and 1 is the identity element for multiplication. The history of the introduction of 0 is interesting and enlightening. Approximating the value of π and e can be done in various ways. Buffon devised a way of approximating π by probability methods. See Eves (31, p. 88). The value of π can be approximated by tossing a coin or a needle. The value of e can be approximated by spinning a needle or by randomly drawing numbers or marbles from a bag. The mathematical theory behind these methods is rather interesting. ϕ, associated with the golden section, occurs not only in mathematics but also with phenomena in nature. The various types of infinities that provide information on these topics are included in Ghyka (40); Hobson (55); Huntley (59); Peter (98); Rucker (106); Sondheimer (125); and Spencer (126).

Several lists of mathematics projects are available. One such list, "Interesting Projects in Mathematics," was prepared by Marc Swadener and Jerry McIntosh. This unpublished list is available from Marc Swadener, School of Education, University of Colorado, Boulder, CO 80302, or Jerry McIntosh, School of Education, Indiana University, Bloomington, IN 47401.

A mimeographed guide for geometry fairs, including suggested projects, possible references, and leading questions for some specific projects, has been developed by Rheta Rubenstein, Mathematics Department, Renaissance High School, Detroit Public Schools, Detroit, MI 48202. The format for the projects includes visual, written, and oral components.

Leading Questions

Another way for a teacher to generate topics for projects is to pose leading questions like the ones below, to which the answers are known. One hopes that these questions will stimulate students' curiosity and lead them to further study in theoretical and applied mathematics.

1. What are curves of constant breadth?
2. Can a set contain itself as an element?
3. Why are there only five Platonic solids?
4. What curves can be drawn in a single stroke without retracing any line or lifting the pencil from the paper?
5. How many mail routes must the post office maintain in order to serve all parts of a city without having more than one route in any section?
6. What are some of the geometric constructions that can be made if tools other than the straightedge and compass are permitted?
7. Which are more numerous, the points of a line segment or the points of the surface of a square?
8. What geometric curve is used for the teeth on cogwheels to provide exceptional smoothness?
9. What are some applications of symmetry in nature?
10. What are the different kinds of networks, and what are the uses of each?
11. What is meant by the statement, "The average height of boys today is five centimeters more than that of boys a generation ago"?
12. Are there more whole numbers than even numbers?
13. What are some applications of mathematics to nature?
14. The sections of a cone are called conic sections; what are some comparable sections of other geometric solids?
15. Can a segment of length $2+\sqrt{7}$ be constructed using a straightedge and compass only?
16. What is the triangle of shortest perimeter that can be inscribed in a given triangle?

Some Unsolved Problems

It is easy for someone who has not studied mathematics deeply to conclude that mathematics is one discipline that has answers for all its problems. Further study, however, shows that many problems are yet to be solved. Some are much older than others, but as old problems are solved, new ones arise. Previously unsolved problems have been solved by the application of new methods or new approaches. The four-color problem remains unsolved using traditional approaches, but many mathematicians believe it has been solved recently using the computer. The following list of unsolved problems might inspire some interesting projects.

1. The equation $a^n + b^n = c^n$ for any positive integral n for $n > 2$ is known as Fermat's last "theorem." (See E. T. Bell's *The Last Problem* [Simon & Schuster, 1961].)
2. Goldbach conjectured that every even number $N \geq 6$ is the sum of two odd primes and that every odd number $N \geq 9$ is the sum of three odd primes.
3. Can an odd integer be perfect? How many perfect numbers are there?
4. Can a simple formula be found such that when a prime number is given, the next larger prime number can be found?
5. Are there infinitely many twin primes?
6. How shall a traveling sales representative plan an itinerary for a trip to an arbitrary number of cities on a map to make the trip as short as possible?
7. Is there a mathematical method to analyze the turbulence of the traffic on a four-lane highway?
8. Can a machine be designed to precisely duplicate the human thought process?
9. Is there a checkable proof of the four-color conjecture?
10. What are some straightforward methods for solving certain nonlinear differential equations that describe the aerodynamic shock waves produced when an airplane accelerates through the speed of sound?
11. When one uses the computer, what are some ways to forestall grossly inaccurate results or situations in which the computer produces no answer at all?
12. How can computer algorithms be proved valid?
13. Can one give a consistent mathematical meaning to the common intuitive notion of straightness?

In 1900 David Hilbert delivered a lecture before the International Congress of Mathematicians in Paris in which he listed and discussed twenty-three unsolved mathematics problems. Since that time many of these have been solved. Another reference for unsolved problems is the textbook *Unsolved Problems in Number Theory* by Guy (48).

3

General Topics

THE FOLLOWING TOPICS are examples of the way in which a particular project can be chosen and developed. Each topic is discussed briefly, suggestions and questions are given to stimulate further thought and study on some particular aspect, and a selected list of references is provided.

High School

Group Theory

A group is one of the simplest and most important mathematical systems, and the basic concept can be understood by many high school students. The concept of groups is fundamental in crystallography, cryptanalysis, and quantum mechanics. Models depicting groups of movements for the triangle, the circle, and the square can easily be constructed of balsa sticks and can even be made to show three dimensions. Applications can also be made to elementary trigonometry. Multiplication tables for these groups of movements can be constructed of balsa wood or drawn on paper.

How does the algebraic concept of groups help to unify different branches of mathematics? What is a loop? How is the theory of braids related to group theory? Does knot theory have a similar relationship? Does modular arithmetic have a connection with groups? Do certain Boolean algebras that form a group under specified operations have a connection? What is the fundamental connection between geometry and modern abstract algebra?

Burton (15)
Durbin (26)
Foulis (32)
Fraleigh (35)
Gilbert (41)
Herstein (51)
Hess (174)
NCTM (155, 160)

Binomial Theorem

This topic is often studied in high school mathematics classes. It can be used to compute compound interest and depreciation and to approximate the roots of numbers, as well as for many other purposes.

Models can be made of the extension of the expansions of a binomial to a positive integral power to the fourth dimension and higher. Balsa sticks, wire, or a solid piece of wood cut properly can be used to make such models. What is a tesseract? What is Pascal's triangle? Can this figure be constructed with a ruler and a compass? What is the compound interest function? How does this function apply to such diverse phenomena as the growth of timber, the increase of bacteria, the surface healing of wounds, the decomposition of radium, or the loss of heat or electrical energy? What uses are made of the binomial model in statistics? Can this theorem be used in studying blood types or Mendel's laws? Does the theorem hold for polynomials? What theorem dealing with complex numbers uses the binomial theorem? Is there any connection between combinations and the binomial theorem?

Courant (21)
Hunkins (58)
Jacobs (61)
Manvel (81)
Mott (89)
NCTM (160)
Rosen (104)
Tucker (134)

The Golden Section

If a line is divided into two segments so that one section is the mean proportional between the whole line and the other segment, the line is said to be divided into extreme and mean ratios. The Greeks considered this proportion to have mystical significance and called it the *divine proportion.* Today it is known as the golden section. A rectangle formed by using the longer part of the golden section as its length and the shorter part as its width is considered to have the most eye-pleasing proportions. This construction is used in inscribing a regular decagon in a given circle.

What is the Fibonacci sequence? What relationship does this have to the plant world? The animal world? Do any of the Platonic solids exemplify the golden section? Are Pascal's triangle and the golden section related?

Cundy (23)
Garland (38)
Ghyka (40)
Huntley (59)
Jacobs (61)
Manuel (183)
NCTM (160)
Schwartzman (189)
Smart (118)

Some Geometric Transformations

A transformation on the plane is a correspondence, rule, or mapping from the set of points in the plane onto the plane. Rotations about a point, translations, reflections about a line, and glide reflections are examples of transformations, often called *rigid motions*. Noncongruent similar figures are examples of nonrigid transformations. In circular inversion, transformations include all the special transformations. The use of matrices in considering transformations serves to integrate algebra and geometry. What properties of plane figures are invariant under each transformation? How is the symmetry of a figure associated with reflections? What applications are made of these transformations in physics? From a theoretical standpoint, what is important about Peaucellier's cell? What transformation does the pantograph illustrate? Can circular inversion be applied when some conic other than a circle is used? What is meant by the statement, "A set of transformations is called a group"? What are matrices, and how are they helpful in work with transformations? Are poles and polars used in a type of transformation?

Black (12)
Brieske (169)
Collier (18)
Courant (21)
Gans (36)
Martin (84)
Mira Activities (85)
Moise (86)
Newman (90)
Smart (118)
Thompson (132, 191)

Computers

The development of high speed computers has spurred a growth of interest in mathematics. The computer is an outstanding example of the application of mathematics to science. As a result of computers, old fields of mathematics and science have been reexamined and new fields created. Changes are so rapid that constant use must be made of periodicals to keep abreast with the current status of both hardware and software.

What, essentially, is a floating-point number used in high-speed computers? How is number theory used? Why is Boolean algebra so applicable to switching circuits? What old mathematical subjects have become important with the advent of computers? How can the computer be adapted to study social and economic problems? What is the significance of the fact that the computer has been used to obtain π and e to thousands of decimal places? How do computers compose music and translate languages? What effects does automation have on human society? Can the computer learn or create? How is the computer used to aid in the solution of some of the unsolved problems in mathematics? What are some of the problems involved in the limited number of digits stored during computations? How can a computer aid the process of information retrieval? How are the ideas of set theory, topology, and symbolic logic being applied to computers? Why has the computer helped probability in mathematics to become a very useful tool for simulation and modeling? How can a computer be used to generate conjectures that eventually might be proved as theorems?

Bitter (11)
Elgarten (27)
Evans (29)
Gnanadesikan (43)
Kimberling (179)
Kosniowski (70)
Mandelbrot (80)
Masalski (184)
Moreau (87)
O'Shea (91)
Ritchie (102)
Schank (109)
Shurkin (117)
Smith (119)
Snover (121)
Spencer (126, 127)
Williams (143)

Finite Geometries

One of the most famous examples of a system of mathematical logic is Euclidean geometry. However, many simpler systems display the interrelationship of the several parts without the many complications of Euclidean geometry. Such geometries are often referred to as *finite* or *miniature* geometries. Since a finite geometry requires fewer postulates than Euclidean geometry, it provides a simpler example of a logical structure. It is usually easier in a finite system to make a study of the consistency, completeness, and independence of a particular set of postulates.

What is the advantage of duality in a geometric system? Do all finite geometries have duality? Are there parallel lines in all finite geometries? How does one test a set of postulates for consistency, completeness, and independence? What is meant by isomorphism? What practical applications of finite geometries can be made to airplane flights, the formation of committees, the planting of trees, or statistical experiments? Can one have a geometry of number triples? Under what geometry are the finite geometries classified?

Runion (107)
Salisbury (188)
Smart (118)

Modular Systems

Some frequently used systems violate many of the familiar rules of mathematics. Some of these are calendar numbers, clock arithmetic, finite arithmetic, and modular arithmetic. A modular system is a simplified example of a finite system that can be isolated and carefully studied. Such systems can be extended as far as interest and time permit.

Are fractions and negative numbers needed in modular systems? What is meant by "divisors of zero"? Do all elements of modular systems have reciprocals? What laws hold in modular systems that do not hold in the real number systems? Can one prove or disprove a theorem by testing all the cases in the real number system? Is this equally true in modular systems? If it is assumed that the usual definition for square root is used, do all numbers in a modulo 6 system have a single square root? Can this be generalized? Does the degree of the equation in modular arithmetic give a clue to the number of solutions expected? Are there actual applications of modular arithmetic in everyday life? Does modular arithmetic have an application in trigonometry? Can modular arithmetic be used to prove the theorems for casting out nines or elevens? Is there a theoretical counterpart in number congruences to the ordinary arithmetic logarithm? How can modulo 9 be used to test whether a natural number is a perfect square? Is modulo 10 used in the base-ten system in the four fundamental operations? How is modular arithmetic applicable to coding messages? Can the concept of congruence be used to determine if any integer in base ten is divisible by 7? Can it be shown that congruence modulo 0 (zero) is ordinary equality?

Courant (21)
Durbin (26)
Gilbert (41)
Schroeder (110)
Somerville (124)
Spencer (126)

Circles, Lines, Points, and Triangles in Modern Geometry

Some of the so-called modern geometries require only high school algebra and plane geometry as a background. The geometry of the circle and the triangle, developed during the nineteenth century, has many special lines and points associated with it. A display of some of the surprising properties of these unique lines and points would be quite interesting. The proof of the construction and any special applications or items of interest connected with the figure should be included as a part of the exhibit. Some of the names associated with the points are Brianchon, Brocard, Ceva, Gergonne, Lemoine, Miquel, Nagel, Steiner, and Tarry. Additional terms associated with circles, lines, and triangles are Euler, the nine-point circle, Pascal's line, the pedal triangle, Simson's line, Speiker's circle, and Tucker's circle.

What theorems in plane geometry are easily proved using Ceva's theorem or Menelaus's theorem? What are the trigonometric forms of these theorems? Can a physical interpretation be made of any of the special points or lines?

Courant (21)
Jacobs (61)
Salisbury (188)
Schwartz (112)
Smart (118)

Inequalities

Inequalities are now included in high school mathematics. This is appropriate because both in nature and in human society inequalities are the rule and equalities are the exception. Such problems as load-carrying capacity, the serviceable life of a product, and sales volume involve applications of inequalities. Inequalities are also of importance in such recent developments as game theory and linear programming.

How are inequalities used in the Dedekind-cut concept of irrational numbers? What are the restrictions on the important inequality $(1 + p)^n \geq np$? What restriction or modification must be made in order to talk about the inequality of complex numbers? How are problems about maxima and minima applications of inequalities? How are inequalities used in modern mathematical economics, game theory, and linear programming?

Beckenbach (7)
Burk (170)
Coppins (20)
Courant (21)
Wiener (193)

i, e, and π

These important mathematical constants have long been objects of study by mathematicians. Both π and e can be expressed as series, and both can be obtained experimentally. π is the ratio of the circumference of a circle to the diameter of the circle; e is sometimes known as Euler's number. Both constants have been calculated to a great number of decimal places with the aid of the computer. Buffon devised a way of approximating π by probability methods. π can be approximated by tossing a coin or a needle, and e can be approximated by spinning a needle. See Eves (31, p. 88).

What is the problem entitled "The Squaring of a Circle"? What constructions in plane geometry are closely associated with π? What constant is in the "normal curve" function? What function equals its own rate of change? What is the curve assumed by a freely hanging chain draped over two pegs? What trigonometric functions use the constant e? What is the relationship between e and natural logarithms? What is the "snowball law"? What is the compound interest law? In what kinds of application problems is e used? How can e and π be approximated statistically? How can continued fractions be used to approximate these two constants? What methods were used to compute e and π before the advent of high-speed computers? Are e and π algebraic numbers?

What is i (as used in mathematics)? What formula expresses a relationship between e, i, and π?

Asimov (5)
Courant (21)
Crossley (22)
Cundy (23)
Durbin (26)
Eves (30)
Ghyka (40)
Hobson (55)
Wells (138)

Non-Euclidean Geometry

The term *non-Euclidean* is used to describe systems of geometry that differ from Euclid's by at least one postulate, often the parallel postulate. Many attempts were made to prove the parallel postulate of Euclid with the aid of the other postulates and axioms but eventually the parallel postulate was proven to be independent. Elliptic and hyperbolic geometries developed from these investigations during the nineteenth century.

Taxi-cab geometry is a simple, non-Euclidean geometry in which the side-angle-side congruence statement for triangles does not hold. This geometry uses the coordinate plane, builds on the basic ideas of Euclidean geometry, and has many practical applications. Taxi-cab geometry provides an excellent medium for individual exploration and creativity.

What properties of postulates do mathematicians study when setting up a mathematical system? What theorems of Euclidean geometry hold in non-Euclidean geometry? What is a geodesic figure? Is a straight line always the shortest distance between two points? How is a pseudosphere generated, and what are some of its properties? What is the relationship between a catenary and a tractrix? What is meant by elliptic, hyperbolic, and parabolic geometries? How are these geometries useful in explaining temperature changes in the universe, the theory of relativity, optics, atomic physics, or the general theory of wave proportion? What do models of elliptic and hyperbolic geometries look like? Is trigonometry associated with each of these geometries?

Courant (21)
Eves (30)
Greenberg (46)
Kline (67)
Krause (73)
Martin (84)
Runion (107)

Time Curves

The path traced by a fixed point on the circumference of a wheel as it rolls, without slipping, on a fixed straight line has many interesting properties. It is at once the path down which a particle will fall from one given point to another in the shortest time and the path down which a particle will fall in equal time no matter what the starting point may be. The semicubical parabola is another curve associated with time. It is a curve such that in any two equal intervals of time, a particle falls equal vertical distances.

What are the properties of other curves closely allied to the cycloid? What important study of mathematics was given impetus by work on the brachistochrone problem? Which one of the time curves has been considered a very important discovery for clock makers? With what other phenomenon in physics did John Bernoulli associate the brachistochrone problem in order to solve the problem? Is it possible for parts of a train to be traveling in opposite directions? What other time curves are there?

Cook (19)
Courant (21)
Cundy (23)
Yates (147)

Four-Dimensional Geometry

We live in a three-dimensional world, but can you visualize a fourth dimension? This task would be similar to someone living in two dimensions trying to visualize a three-dimensional world, a concept addressed in *Flatland,* a book by Edwin A. Abbott. The idea of a fourth dimension has been developed mathematically since 1847 and can be approached from any of three perspectives: that of choosing a fundamental element, that of analytic geometry, and that of synthetic geometry.

What is the history of the nineteenth-century development of the concept of four dimensions? What use did Einstein make of the concept in the early part of the twentieth century? How can algebra be used to represent four or more dimensions? What are some differences between the geometry of three dimensions and that of four dimensions? What is a hypercube? A tesseract? What are some applications that use the concept of the fourth dimension?

Abbott (1)
Courant (21)
Henderson (49)
Hess (52, 175, 176)
Hinton (54)
Marr (83)

Topology

Topology is a study of properties that continue to hold under change or transformation. It is an area of mathematics in which triangles and circles are considered equivalent, where a surface can have exactly one side and a donut is equivalent to a coffee cup. Euclidean geometry is just a special topic area within topology.

What connection exists between the famous Königsberg bridge problem, the four-color problem, the Möbius strip, and Klein's bottle? What happens to a Möbius strip if it is cut down the middle? Will the results be the same if some other method of cutting is tried? What are some tricks and puzzles based on topology?

Armstrong (4)
Barnette (6)
Collier (18)
Courant (21)
Jacobs (61)
Kline (67)
Poggi (186)
Smart (118)

Hand-held Calculators

Most people, including students at all levels, own hand-held calculators. The reduction in their cost and the great variety of calculators available have contributed to this phenomenon. The potential contribution of this instrument as a valuable instructional aid is well known. Students are urged to use calculators in imaginative ways and for enrichment. Many persons and groups continue to study ways to use calculators in teaching mathematics, as well as to apply them to new problems. Puzzles and games have been devised that are solved with a calculator. For example, how can a calculator be used to find the remainder when 5999 is divided by 7?

What is the historical development of calculators from modern times to the present? What are some variations of the game of nim using the calculator? Is it possible to change fractions to decimals and then reverse the process? Are there other such procedures that are hard or impossible to do on a calculator? Are there calculators for the blind, or talking calculators? What are the effects of rounding off or truncating on computations carried out with a calculator? What is the impact of calculators in everyday life? Can two-digit numbers be classified relative to the number of additions needed before the sum is a palindromic number? How can a graphics calculator be used to find the solutions to an equation or a system of equations?

Waits (136)
Williams, D. E. (142)
Williams, M. R. (143)
Zehna (148)

Numerical Analysis

Numerical analysis is a branch of mathematics in which computational algorithms are used to find numerical solutions to problems. Engineers and scientists sometimes create equations that model physical situations but are difficult or impossible to solve by hand. The computer is the primary tool used to implement numerical algorithms.

What algorithms can be used with the computer to find roots of polynomials or solutions to any equation? How can graphing help in determining numerical answers? What is an iterative method? Is recursion involved? Why is the concept of convergence of a sequence important? How can someone assess the accuracy of an approximate solution when the exact answer is not known? What methods may be used to solve systems of linear equations? Do the methods work for nonlinear systems?

Gerald (39)
Hua (56)
Jain (63)
Johnson (64)
Maron (82)
Morris (88)
Oldknow (92)
Vandergraft (135)

Logic and Proof

Some mathematicians define proof as a valid argument, based on a system of logic that convinces one's peers. To get to the point where proof is needed, we must have a conjecture to work with and a set of rules on how to proceed. In all areas of mathematics, definitions, postulates, and theorems provide the foundation for new ideas as conjectures. Before these conjectures can be accepted as theorems, they must be proved by showing that they have been derived from previously accepted postulates and theorems.

What attempts have previously been made to develop logical systems for mathematical reasoning? Is "truth versus falsity" the same idea as "valid versus invalid"? What are some examples of invalid reasoning that are persuasive in politics? What is a tautology? Give three examples of paradoxes. What are various types of direct and indirect proof? What attempts have been made to use the computer to prove or disprove conjectures?

Agostini (2)
Bunch (14)
Foulis (32)
Gardner (37)
Horak (178)
Solow (122;123)
Stolyar (130)
Wilder (141)
Wylie (146)

Probability and Statistics

Graphs, data, and predictions are encountered when one reads current periodicals and newspapers. Industry and government agencies formulate budgets, quotas, and general policy based on theories from probability and statistics. Some people describe the evolving field of statistics as more of an art than a science.

What does the law of large numbers imply about rolling a fair die? In the game of Keno, are the expected winnings greater two numbers are played or ten numbers? How can the results of a presidential election be predicted when fewer than 10 percent of the votes have been counted? What is meant by the statement that the "results are statistically significant"? How does the concept of *confidence interval* directly relate probability and statistics? Can statistics lie? Can data provided by simulating a real-world problem assist in solving the problem? What distributions, other than the normal curve, have useful applications?

Gnanadesikan (43)
Huck (57)
Huntsberger (60)
Jaffe (62)
Kirk (66)
Kooker (69)
Kotz (72)
Landwehr (75)
Lindgren (76)
NCTM (162)
Newman (90)
Packel (95)
Rector (187)
Robinson (103)
Zehna (148)
Zeisel (149)

Junior High School

Computing Methods, Calculating Instruments, and Measuring Devices

Junior high school students can construct and use many interesting calculating instruments and measuring devices. Although their construction and use are important and worthwhile, the mathematical principle on which the device or instrument depends is even more important. If the principle can be applied in a new situation or put to a new use, the resulting instrument becomes even more worthwhile. The various devices and instruments have interesting histories. Some common ones are the abacus, Napier's bones or rods, compass, ruler, calculator, computer, vernier calipers, and the protractor.

What other mathematical devices did Napier invent? Are the rods still used? Where is the abacus still used? Is the calculator faster than the abacus? What kinds of instruments can be devised to solve simultaneous linear equations? What uses can be made of nomographic models? Who invented logarithms? What are the scratch, doubling and summing, and lattice methods of multiplication? What is finger multiplication and why does it work? What contributions did Babbitt, Von Neumann, Kemeny, and Papert make with regard to computers?

Goldberg (44)
Good (171)
NCTM (153;155;164)
Schwartz (111)
Smith (119;120)
Spencer (126)

Numeration Systems

Numeration systems can well be a fertile source of topics for projects in mathematics for junior high school students. The rudiments of many of the ideas may already be familiar. The theory behind them as well as the application should intrigue any student who is interested in mathematics.

What number systems (bases) are used in computers? What is mean by "casting out" nines or elevens? Can numbers for other bases be found that "cast out" in a manner similar to nine and eleven in base ten? What two fundamental principles are used in writing numerals in the Hindu-Arabic system? What are some other numeration systems, and what are the principles used in writing numerals in these systems? Can the usual arithmetic operations carried out in base ten be done in any base? What are some devices used for computing in the base-ten numeration system? Can these be made to function in any system? What is meant by structure in arithmetic? What are sets, and how can they be applied to arithmetic? What are some ancient methods of recording results of counting? What properties of the base-ten numeration system are applied in "pick a number" tricks? What is the Euclid algorithm? What is the largest number? In true-false questions if an answer is either right or wrong and the score is the number right, what numeration system is being used? Is this the same system that is used in nim or in using a set of cards to tell a person's age? What is the "Russian peasant" method of multiplication? Can negative numbers be used for bases? Can fractions be used? What is a symmetric numeration system?

Arcvi (166)
Billstein (10)
Henry (50)
NCTM (153)
Vest (192)

Statistics and Probability

Because the data used in statistics must be obtained by measuring, recording, or weighing various objects and phenomena, measurement and statistics are treated together here. Each could well be the fundamental part of a project or study. Both concepts are used in everyday life. Elementary concepts of statistics are included in most school mathematics textbooks. All biological and physical phenomena lend themselves to measurement and the subsequent application of statistics. A study involving one or both of these concepts can be somewhat elementary, or it can be quite complex and extend over several years. A number of particular suggestions for the study of these concepts are included in materials produced by the Quantitative Literacy Project, a joint effort by the American Statistical Association and the National Council of Teachers of Mathematics. See chapter 4.

What is probability? What are some misuses of statistics? What are some practical applications of probability? What are some measures of central tendency? Of dispersion? Into what kind of statistical pattern do errors of measurement fall? What is the foundation of statistics? What is inferential statistics? How are the results from the four arithmetic operations affected when approximate data are used? How is a survey conducted? Can the results be used to make predictions about a large population? How can scatterplots, box-and-whisker plots, and stem-and-leaf plots be used to represent data?

Gudder (47)
Landwehr (74)
NCTM (162, 163)
Newman (90)
Niman (185)
Pearson (96)
Runyon (108)
Smith, J. K. (120)
Stein (129)

Paper Folding

Geometric figures have been constructed within the restrictions of the straightedge and compass for centuries, but other methods were also used. Paper folding and creasing can be used, with certain assumptions, to make any construction that can be made with the traditional tools. Paper folding can be effectively used to develop or extend a project in mathematics at the junior high school level.

What assumptions are made when paper folding is used to construct geometric figures? What polygons and polyhedrons can be made by paper folding? How can paper folding be used to show certain mathematical fallacies? How can paper folding be extended to "knotting"? How can the method of paper folding be extended to hexaflexagons? Does this method tend to unify algebra and geometry? Can this method be used to find the area of a triangle?

Bohen (168)
Cundy (23)
Davidson (25)
Olson (93)
Row (105)
Wenninger (139)

Curve Stitching

Curves can be constructed by using a straightedge, paper folding, or string to create straight lines that form curves.

How can curves be formed by straight lines? What curves can be formed by stitching? Can curve stitching be used to make three-dimensional models of curves? What are some applications of these curves? What is the envelope of a curve? What is a pencil of lines?

Cundy (23)	Pohl (100)
Lyng (79)	Seymour (116)
NCTM (160)	Winter (145)

Optical Illusions

Optical illusions are used in many ways. Illusions are an essential tool in some types of art work. Special mirrors can produce optical illusions that make us laugh. Occasionally, illusions lead us to erroneous conjectures such as when line segments appear to be of different lengths but are congruent.

What are some types of optical illusions? Why do these patterns cause illusions? Can optical illusions be extended to three dimensions? What tricks depend on optical illusions? How are optical illusions used in camouflaging, advertising, dress design, and architecture?

Armstrong (4)	Jacobs (61)
Ernst (28)	Luckiesh (78)
Gregory (172)	Smith, J. K. (120)

The Metric System

The metric system, a system of weights and measures based on the decimal system, was adopted in France in 1799. The metric system is used throughout the world and is the official system of measure in the United States although the English system is still predominant. The United States is the only developed country in the world where the metric system is not in widespread use.

What is the history of the metric system? What is SI? Is there an aid for learning the metric system? What are some metric prefixes not commonly used? What are some strong arguments in favor of replacing the English system with the metric system? What are some arguments against it?

Alexander (165)	NCTM (158;159)
Gilbert (42)	

Motion Geometry

Many people believe that the most useful everyday geometry involves slides, flips, and turns. In 1872, at the age of 23, Felix Klein, a professor at the University of Erlangen, developed these ideas. He defined geometry as the study of those properties of figures that do not change when a figure is transformed into another figure by flips, slides, or turns. This geometry is called motion or transformation geometry.

What are some everyday examples of motion geometry? How can motions be duplicated using the compass, straightedge, or paper tracing? Can a slide or turn be duplicated by just using reflections? How is a Mira used in motion geometry? How can the computer language called Logo and the movement of the turtle teach us about motions?

Bidwell (167)
Billstein (9;10)
Black (12)
Collier (18)
Kenney (65)
Mira Activities (85)
Willson (144)

4

Resources

Annotated Bibliography of Mathematics Books and Periodicals

AN EXTENSIVE LIBRARY of mathematics books and periodicals is a fertile source of ideas for mathematics projects. A number of lists of mathematics books suitable for the junior and senior high school library are available. A careful selection of materials from the lists given here will provide the nucleus for a good library in mathematics.

Books

The following books listed are *primary source* books for projects:

Goodman, Harvey, et al. *A Guide to Establishing a Science/Mathematics Research Program in High School.* New York City Board of Education, 1982.

Schaaf, William L. *A Bibliography of Recreational Mathematics.* Vols. 1–4. NCTM, 1970, 1970, 1973, 1978.

Volumes 1, 2, and 3 list the best of the literature up to 1973. Volume 4 updates the material; lists more than 2600 references, most published since 1972; and includes a supplementary glossary of over 200 terms not given in volume 3.

———. *The High School Mathematics Library.* 8th ed. NCTM, 1987.

The eighth edition of this bibliography, reflecting new trends and needs, has wider coverage of computers and calculators, computer recreations, programming automation, expository and recreational mathematics, and professional books for teachers.

Posamentier, Alfred S., and J. Stepelman. *Teaching Secondary School Mathematics.* 2d.ed. Merrill, 1986.

The book has two parts: Part 1 is a guide to daily teaching, and part 2 is a unique section containing a wealth of enrichment ideas for both junior and senior high students. The book includes an extensive bibliography to aid in project searches.

Projects to Enrich School Mathematics, Levels 2 & 3. 2d ed. Edited by Leroy C. Sachs. NCTM, 1988.

These two independent collections offer challenging projects for students at all levels. Each unit contains enrichment materials with hints, drawings, references and ideas for further investigation. Sample units include "Topology," "Pi and Its History," "The Mathematics of Flight," and "Paradoxes in Mathematics."

Thomas, David A. *Math Projects for Young Scientists.* Franklin Watts, 1988.

A book intended for self-directed students as well as their mentors. Background information for a number of mathematics and science projects is given, as well as hints for constructing exhibits.

Topics for Mathematics Clubs. 2d ed. Edited by Leroy C. Dalton and Henry C. Snyder. NCTM, 1983.

Stimulates interest in mathematical investigation through exciting topics not usually discussed in the classroom. Bibliographies suggest further reading.

The following books are good *general resources* for projects:

Activities from the "Mathematics Teacher". Edited by Evan M. Maletsky and Christian R. Hirsch. NCTM, 1981.

A useful source of discovery lessons, laboratory experiences, mathematical games and puzzles, and model constructions for use in grades 7 through 12. This convenient compilation includes the topics of computational skills, calculators, geometry, measurement, and problem solving.

Applications in School Mathematics. 1979 Yearbook. NCTM, 1979.

Many practical and varied applications of mathematics in the liberal arts, business and the sciences; annotated bibliography.

Campbell, Douglas, and John Higgins. *Mathematics: People, Problems, Results.* Vols. 1, 2, & 3. Wadsworth, 1984.

Extensive collection of readings about mathematics: how it developed and philosophical and psychological implications.

Chips from the Mathematical Log and *More Chips from the Mathematical Log*. Edited by Josephine P. Andree. Mu Alpha Theta, 1966, 1970.

Computers in Mathematics Education. 1984 Yearbook. NCTM, 1984.

Answers questions dealing with computers in the classroom; computers are discussed from the standpoint of a diagnostic and problem solving tool with programming used as a means of teaching mathematics; good bibliography for further reading.

Courant, Richard, and Herbert Robbins. *What Is Mathematics? An Elementary Approach to Ideas and Methods*. Oxford, 1941.

Originally written for mature readers, now a classic; excellent introductory ideas on number systems, geometric construction, number theory, topology, and postulational systems.

Cundy, Henry Martyn, and A. P. Rollett. *Mathematical Models*. Oxford, 1952.

An excellent discussion of various mathematical models.

Enrichment Mathematics for High School. Twenty-eighth Yearbook. NCTM, 1963.

For students in high school and beyond. Extensive bibliography.

Fox, Lynn H., et al., eds. *Women and the Mathematical Mystique*. Hopkins, 1980.

Extension of talks and ideas presented at the 1976 AAAS Symposium entitled "Women and Mathematics."

Halmos, P. R. *Selecta: Expository Writing*. Edited by D. E. Sarason and Leonard Gillman. Springer-Verlag, 1983.

Stimulating; for mature students and teachers.

Huntley, H. E. *The Divine Proportion: A Study in Mathematical Beauty*. Dover, 1970.

Some of the topics treated are patterns, Fibonacci numbers, Pascal's triangle, golden numbers, and the divine proportion.

The Ideas of Algebra, K–12. 1988 Yearbook. NCTM, 1988.

Depicts algebra as a living, growing curricular field composed of concepts that should be understood rather than a subject just considered a "bag of tricks." Topics include equations and expressions, word problems, and the use of technology in the algebra classroom as well as tested ideas for teaching algebra.

Instructional Aids in Mathematics. Thirty-fourth Yearbook. NCTM, 1973.

A guide to instructional aids. The many illustrations and ideas for construction make this a valuable book for anyone making an exhibit.

Kastner, Bernice. *Applications of Secondary School Mathematics*. NCTM, 1978.

Helps teachers answer the recurring question "What's it good for?" with responses the student will understand and find interesting and important. Contains real applications from such fields as physics, chemistry, biology, and economics, with which the mathematics teacher sometimes has little experience.

Kline, Morris, ed. *Mathematics: An Introduction to Its Spirit and Use*. Freeman, 1979.

Presents readings from *Scientific American* and a very readable, broad introduction to mathematical ideas and applications; the history of mathematical thought from number theory, algebra, geometry, and statistics, to symbolic logic and computers; provides ideas and applications.

Kline, Morris. *Mathematics: The Loss of Certainty.* Oxford, 1980.

A history of the relationship between mathematics and science and of the development of the philosophy and foundations of mathematics from antiquity to the present; requires some mathematical maturity, but nicely written.

Learning and Teaching Geometry, K–12. 1987 Yearbook. NCTM, 1987.

Geometry depicted in problem solving and its applications; activities; blending geometry with other areas of mathematics; how students learn geometry including van Hiele levels.

The Man-Made World. Engineering Concepts Curriculum Project. McGraw, 1971.

Three ideas have guided the authors: (1) learning should be fun, (2) the subject matter should be relevant, and (3) science should be easy. This is not a course on the scientific method. Mathematical problem solving is secondary to the process of logical thinking. The book presents a series of significant, current problems in which the concepts provide understanding, and it gives down-to-earth applications of mathematics to everyday situations.

Mathematical Association of America and National Council of Teachers of Mathematics. *A Sourcebook of Applications of School Mathematics.* NCTM, 1980.

Contains more than 360 pages of reasonable, realistic mathematical problems at the secondary school level that have direct applications to many varied activities. Emphasizes the wide applications of mathematics, gives students a more balanced appreciation of the place of mathematics in our civilization, and helps them to think mathematically in everyday life. Answers are provided.

Mathematics in the Modern World. Freeman, 1968.

Readings from the *Scientific American* with introductions by Morris Kline. These articles, provocative and authoritative, cover a wide range of mathematical topics.

Multi-Sensory Aids in the Teaching of Mathematics. Eighteenth Yearbook. NCTM, 1945.

An excellent source of ideas for exhibits and methods for constructing them.

Organizing for Mathematics Instruction. 1977 Yearbook. NCTM, 1977.

Provides ideas for organizing independent learning units to supplement regular class instruction.

Perl, Teri. *Math Equals: Biographies of Women Mathematicians and Related Activities.* Addison, 1987.

Discusses the personal lives and work of nine famous female mathematicians who overcame obstacles to make significant contributions.

Pólya, George. *How to Solve It.* Princeton, 1956.

A four-step problem-solving procedure is spelled out carefully and clearly.

———. *Mathematical Discovery: On Understanding, Learning, and Teaching Problem Solving.* Wiley, 1981.

The author is a master at leading readers to guess answers and to discover mathematics.

Problem Solving in School Mathematics. 1980 Yearbook. Edited by Stephen Krulik. NCTM, 1980.

Contains more than twenty useful articles with problems, examples, illustrations, and ideas to be used in the classroom, along with a lengthy annotated bibliography of other resources. Helps teachers in the important effort to teach problem solving—a basic skill that students must take with them throughout their lives.

Providing Opportunities for the Mathematically Gifted, K–12. Edited by Peggy House. NCTM, 1987.

Ideas are based on the premise that the mathematically gifted are a virtually untapped resource who have not been stimulated to reach their full potential.

The Quantitative Literacy Series

Landwehr, James M., and Anne E. Watkins. *Exploring Data.* Seymour, 1986.

Specifics needed to master scatterplots, stem-and-leaf plots, box-and-whisker plots, and line graphs as ways of representing data.

Newman, Claire M., Thomas E. Obremski, and Richard L. Scheaffer. *Exploring Probability.* Seymour, 1986.

Teaches concepts of probability, including sample space, equally likely event, dependent and independent events, and conditional probability.

Gnanadesikan, Mirudulla, Richard L. Scheaffer, and Jim Swift. *The Art and Techniques of Simulation.* Seymour, 1986.

Modeling of real-world events through use of random number tables, spinners, dice, cards, and the computer.

Landwehr, James M., Jim Swift, and Ann E. Watkins. *Exploring Surveys and Information from Samples.* Seymour, 1986.

Discusses the interrelationship between probability and statistics as used in survey results and also addresses the related concept of confidence intervals.

Readings for Enrichment in Secondary School Mathematics. Edited by Max A. Sobel. NCTM, 1988.

A collection of articles previously printed in several NCTM sources; it also contains three original chapters on harmonic mean, rotation matrices and complex numbers, and how computers and calculators perform arithmetic.

Smith, Karl J. *The Nature of Modern Mathematics.* Brooks/Cole, 1980.

Well-written and appropriate exposition for the general reader, students, and teachers.

Solow, Daniel. *How to Read and Do Proofs: An Introduction to Mathematical Thought Processes.* Wiley, 1982.

Specific techniques for beginning and writing proofs: one of the few books devoted to the topic of writing proofs.

Spencer, Donald D. *The Illustrated Computer Dictionary.* Merrill, 1980.

A sourcebook for high school or college students interested in computer science, computer literacy, and data processing.

Teaching Statistics and Probability. 1981 Yearbook. Edited by Albert P. Shulte. NCTM, 1981.

Provides suitable classroom ideas for teaching statistics and probability. Includes graphs, population sampling, generating random digits, games of chance, stem-and-leaf plots, correlation, a computer algorithm, and more.

Periodicals

Periodicals that should be available for use by students include the following:

American Mathematical Monthly. Mathematics Association of America (MAA).

Although this is a college-level magazine, it is a valuable source of ideas and problems.

Arithmetic Teacher. National Council of Teachers of Mathematics (NCTM).

Directed toward content and pedagogy in grades K–8.

Computing Teacher. International Council for Computers in Education (ICCE).

Emphasizes ideas about computers and programming, teaching using computers, teacher education, and the impact of computers on many curricular areas.

The College Mathematics Journal. MAA.

Formerly called *The Two-Year College Mathematics Journal;* mathematics articles appropriate for the upper high school through college mathematics levels.

Journal of Computers in Mathematics and Science Teaching. Association for Computers in Mathematics and Science Teaching.

Mathematical Log. Mu Alpha Theta.

The content is directed to secondary school students.

Mathematics Magazine. MAA.

A college-level magazine that is a source of problems and suggestions.

Mathematics Teacher. NCTM.

This journal is directed toward the content and pedagogy of mathematics at the 7–12 grade levels.

Scientific American.

Every issue devotes space to mathematics. It has served as a fertile source for mathematics projects in the past.

School Science and Mathematics. School Science and Mathematics Association (SSMA).

Problems for high school students are included in this magazine. Applications of mathematics to science may be found here also.

Student Math Notes. NCTM.

Lessons on different mathematical topics for students in grades 7–12.

Articles

The suggestions, comments, and ideas in the following articles are worthwhile for anyone interested in mathematics projects, mathematics fairs, or mathematics clubs.

Bruckheimer, Maxim, and Rina Hershkowitz. "Mathematics Projects in Junior High School." *Mathematics Teacher* 70 (October 1977):573–78.

Dalton, LeRoy C. "A Student-presented Mathematics Club Program—Non-Euclidean Geometries." *Mathematics Teacher* 73 (September 1980):450–51.

Freeman, William, and Diana Krall. "A Mathematics Fair in the Lower Grades." *Arithmetic Teacher* 21 (November 1974):624–28.

Johnson, Donovan. "A Fair for Mathematics with Marathons and a Midway." *School Science and Mathematics* 65 (December 1965):821–24.

Lumpkin, Beatrice. "A Mathematics Club Project from Omar Khayyam." *Mathematics Teacher* 71 (December 1978):740–44.

Publishers: Names and Addresses

Academic
Academic Press, Inc.
465 S. Lincoln Dr.
Troy, MO 63379

Addison-Wesley
Addison-Wesley Publishing Co.
Jacob Way
Reading, MA 01867

Allyn & Bacon
Allyn & Bacon
7 Wells Ave.
Newton, MA 02159

Appleton
Appleton-Century-Crofts
25 Van Zant St.
East Norwalk, CT 06885

Ayer
Ayer Company Publishers
382 Main St.
Salem, NH 03079

Benjamin-Cummings
Benjamin-Cummings Publishing Co.
2727 Sand Hill Rd.
Menlo Park, CA 94025

Birkhäuser
Birkhäuser Boston
380 Green St.
Cambridge, MA 02139

Brooks/Cole
Brooks/Cole Publishing Co.
P.O. Box 1066
Carmichael, CA 95608

Brown
Wm. C. Brown Co. Publishers
2460 Kerper Blvd.
Dubuque, IA 52001

Cambridge
Cambridge University Press
32 E. 57th St.
New York, NY 10022

CBS
CBS Educational & Professional Publishing
383 Madison Ave.
New York, NY 10017

Chelsea
Chelsea Publishing Co., Inc.
15 E. 26th St.
New York, NY 10010

Computer Science
Computer Science Press
11 Taft Ct.
Rockville, MD 20850

Cosine
Cosine Inc.
Box 2017
West Lafayette, IN 47906

Creative Publications
Creative Publications
5005 W. 110th St.
Oaklawn, IL 60453

Dekker
Marcel Dekker
270 Madison Ave.
New York, NY 10016

Dellen
Dellen Publishing Co.
400 Pacific Ave.
San Francisco, CA 94133

Doubleday
Doubleday & Co.
245 Park Ave.
New York, NY 10167

Dover
Dover Publications, Inc.
31 E. 2d St.
Mineola, NY 11501

ERIC
ERIC Clearinghouse for Science, Mathematics and Environmental Education
1200 Chambers Rd.
Room 310
Columbus, OH 43212

Ford Foundation
320 E. 43d St.
New York, NY 10017

Franklin Watts
Franklin Watts Inc.
387 Park Ave., S.
New York, NY 10016

Free Press
The Free Press Inc.
866 Third Ave.
New York, NY 10022

Freeman
W. H. Freeman & Co.
41 Madison Ave.
New York, NY 10010

Ginn
Ginn Custom Publishing
191 Spring St.
Lexington, MA 02173

Goose Pond
Goose Pond Publications
11600 SW Freeway, Suite 179
Houston, TX 77031

Hafner
Hafner Press
866 Third Ave.
New York, NY 10022

Harper & Row
Harper & Row
Key Distribution Center
Dunmore, PA 18512

Heath
D. C. Heath & Co.
125 Spring St.
Lexington, MA 02173

Holden
Holden Day
4432 Telegraph Ave.
Oakland, CA 94609

Holt
Holt, Rinehart & Winston
See CBS.

Houghton Mifflin
Houghton Mifflin Co.
One Beacon St.
Boston, MA 02108

ICTM
Iowa Council of Teachers of Mathematics
University of Northern Iowa
Cedar Falls, IA 50613

Johns Hopkins
Johns Hopkins University Press
Baltimore, MD 21218

Krieger
Robert E. Krieger Publishing Co.
P.O. Box 9542
Melbourne, FL 32902

Little, Brown
Little, Brown & Co. Inc.
34 Beacon St.
Boston, MA 02106

MAA
Mathematical Association of America
1529 18th St., NW
Washington, DC 20036

Macmillan
Macmillan Publishing Co.
866 Third Ave.
New York, NY 10022

McGraw
McGraw-Hill Book Co.
1221 Ave. of the Americas
New York, NY 10020

Merrill
Merrill Publishing Co.
1300 Alum Creek Dr.
Columbus, OH 43216

Midwest
Midwest Publications
P.O. Box 448
Pacific Grove, CA 93950

MIT
MIT Press
28 Carleton St.
Cambridge, MA 02142

Mu Alpha Theta
Box 54
University of Oklahoma
Norman, OK 73069

NCTM
National Council of Teachers of Mathematics
1906 Association Dr.
Reston, VA 22091

New York City Board of Education
131 Livingston St., Room 613
Brooklyn, NY 11201

Norton
W. W. Norton & Co., Inc.
500 Fifth Ave.
New York, NY 10110

Ohio State
Department of Mathematics
Ohio State University
231 W. 18th Ave.
Columbus, OH 43210

Open Court
Open Court Publishing Co.
315 Fifth St.
Peru, IL 61354-2859

Oxford
Oxford University Press
16-00 Pollitt Dr.
Fairlawn, NY 07410

Prentice-Hall
Prentice-Hall, Inc.
Englewood Cliffs, NJ 07632

Princeton
Princeton University Press
41 William St.
Princeton, NJ 08540

Prindle, Weber & Schmidt
Prindle, Weber & Schmidt
20 Park Plaza
Boston, MA 02116-4501

PWS-Kent
PWS-Kent Publishing
Div. of Wadsworth, Inc.
Belmont, CA 94002

Random
Random House
201 E. 50th St.
New York, NY 10022

Saunders
W. B. Saunders Co.
W. Washington Sq.
Philadelphia, PA 19105

Scott
Scott, Foresman/Little
1900 E. Lake Ave.
Glenview, IL 60025

Seymour
Dale Seymour Publications
P.O. Box 10888
Palo Alto, CA 94303

Simon
Simon & Schuster
1230 Ave. of the Americas
New York, NY 10020

Springer-Verlag
Springer-Verlag New York, Inc.
175 Fifth Ave.
New York, NY 10010

Sunburst
Sunburst Communications, Inc.
39 Washington Ave.
Pleasantville, NY 10570

Thomas
Charles C Thomas Publisher
2600 S. First St.
Springfield, IL 62717

Van Nostrand
Van Nostrand Reinhold Co. Inc.
115 Fifth Ave.
New York, NY 10003

Viking Penguin
Viking Penguin Inc.
40 W. 23d St.
New York, NY 10010

Wadsworth
Wadsworth Publishing Co.
10 Davis Dr.
Belmont, CA 94002

Walch
J. Weston Walch, Publisher
321 Valley St.
Portland, ME 04104

Waveland
Waveland Press, Inc.
P.O. Box 400
Prospect Heights, IL 60070

Wiley
John Wiley & Sons
605 Third Ave.
New York, NY 10158

World Scientific
World Scientific Publishing
Company, Inc.
687 Hartwell St.
Teaneck, NJ 07666

Periodicals: Names and Addresses

Arithmetic Teacher
National Council of Teachers of
Mathematics
1906 Association Dr.
Reston, Va 22091

BYTE
BYTE Publications
70 Main St.
Peterborough, NH 03458

College Mathematics Journal
(formerly *The Two-Year College Mathematics Journal*)
Mathematical Association of America
1529 18th St., NW
Washington, DC 20036

The Computing Teacher
International Council for Computers in
Education

University of Oregon
1987 Agate St.
Eugene, OR 97403

Creative Computing Magazine
Creative Computing
P.O. Box 789-M
Morristown, NJ 07960

Games
Games
P.O. Box 10147
Des Moines, IA 50349

Journal of Computers in Mathematics and Science Teaching
Association for Computers in Mathematics and Science Teaching
P.O. Box 60730
Phoenix, AZ 85082

Mathematical Log
Mu Alpha Theta
Box 54
University of Oklahoma
Norman, OK 73069

Mathematics Magazine
Mathematical Association of America
1529 18th St., NW
Washington, DC 20036

Mathematics Teacher
National Council of Teachers of Mathematics
1906 Association Dr.
Reston, VA 22091

Mathematics Teaching
Association of Teachers of Mathematics
Market Street Chambers
Nelson, Lancashire BB97LN, England

Recreational Mathematics
Journal of Recreational Mathematics
Baywood Publishing Co.
120 Marine St.
Farmingdale, NY 11735

School Science and Mathematics
School Science and Mathematics Association
Executive Office
Stright Hall, P.O. Box 1614
Indiana University of Pennsylvania
Indiana, PA 15701

Scientific American
Scientific American
415 Madison Ave.
New York, NY 10017

Note: The NCTM Regional Services Committee has a list of various NCTM Affiliated Groups that publish journals.

5

References

Books and Pamphlets

1. Abbott, Edwin A. *Flatland.* Dover, 1952.
2. Agostini, Franco. *Math and Logic Games.* Seymour, 1986.
3. Albers, Donald J., and Gerald L. Alexanderson. *Mathematical People: Profiles and Interviews.* Birkhäuser, 1985.
4. Armstrong, M. A. *Basic Topology.* Springer-Verlag, 1983.
5. Asimov, Isaac. *Asimov on Numbers.* Doubleday, 1977.
6. Barnette, David. *Map Coloring, Polyhedra and the Four-Color Problem.* MAA, 1983.
7. Beckenbach, Edwin F,. and Richard Bellman. *Inequalities.* Springer-Verlag, 1971.
8. Bezuszka, Stanley, Margaret J. Kenney, and Stephen M. Kokoska. *Applications of Mathematics through Models and Formulas.* Seymour, 1987.
9. Billstein, Rick, Shlomo Libeskind, and Johnny W. Lott. *Apple Logo: Programming and Problem Solving.* Benjamin/Cummings, 1986.
10. ———. *A Problem Solving Approach to Mathematics for Elementary School Teachers.* Benjamin/Cummings, 1987.
11. Bitter, Gary G. *Computers in Today's World.* Wiley, 1984.
12. Black, Howard, and Sandra Black. *Figural Analogies.* Midwest Publications, 1988.
13. Box, Joan. *R. A. Fisher, The Life of a Scientist.* Wiley, 1978.
14. Bunch, Bryan H. *Mathematical Fallacies and Paradoxes.* Van Nostrand, 1982.
15. Burton, David M. *Abstract Algebra.* Brown, 1988.
16. ———. *The History of Mathematics: An Introduction.* Allyn & Bacon, 1985.
17. Christianson, Gale E. *In the Presence of the Creator: Isaac Newton and His Times.* Free Press, 1984.
18. Collier, Patrick C. *Geometry for Teachers.* Waveland, 1984.
19. Cook, Theodore A. *The Curves of Life.* Seymour, 1986.
20. Coppins, Richard J., and Paul M. Umberger. *Applied Finite Mathematics.* Addison-Wesley, 1986.
21. Courant, Richard, and Herbert Robbins. *What Is Mathematics?* Oxford, 1978.
22. Crossley, John N. *The Emergence of Number.* 2d ed. World Scientific, 1987.
23. Cundy, Henry M., and A. P. Rollett. *Mathematical Models.* 2d ed. Oxford, 1961.
24. David, Florence N. *Games, Gods, and Gambling: The Origins and History of Probability and Statistical Ideas from the Earliest Times to the Newtonian Era.* Hafner, 1962.
25. Davidson, Patricia S., and Robert E. Willcutt. *Spatial Problem Solving with Paper Folding and Cutting.* Seymour, 1986.
26. Durbin, John. *Modern Algebra: An Introduction.* 2d ed. Wiley, 1985.
27. Elgarten, Gerald, Alfred Posamentier, and Stephen Moresh. *Using Computers in Mathematics.* Addison-Wesley, 1983.

28. Ernst, Bruno. *Adventures with Impossible Figures*. Seymour, 1987.

29. Evans, Christopher Riche. *The Making of the Micro: A History of the Compute*r. Van Nostrand, 1981.

30. Eves, Howard Whitley. *Great Moments in Mathematics*. Vol. 1: Before 1650. MAA, 1980. Vol. 2: After 1650. MAA, 1981.

31. ———. *An Introduction to the History of Mathematics*. Saunders, 1983.

32. Foulis, David, and Mustafa A. Munem. *After Calculus: Algebra*. Dellen, 1988.

33. Fox, Lynn H. *The Problem of Women and Mathematics: A Report to the Ford Foundation*. Ford Foundation, 1981.

34. ———. *Women and the Mathematical Mystique*. Johns Hopkins, 1984.

35. Fraleigh, John B. *A First Course in Abstract Algebra*. 3d ed. Addison-Wesley, 1982.

36. Gans, David. *Transformations and Geometry*. Appleton, 1969.

37. Gardner, Martin. *Aha! Gotcha: Paradoxes to Puzzle And Delight*. Freeman, 1982.

38. Garland, Trudi H. *Fascinating Fibonaccis*. Seymour, 1988.

39. Gerald, Curtis F., and Patrick O. Wheatley. *Applied Numerical Analysis*. 3d ed. Addison-Wesley, 1984.

40. Ghyka, Matila. *The Geometry of Art and Life*. 2d ed. Dover, 1978.

41. Gilbert, Jimmie, and Linda Gilbert. *Elements of Modern Algebra*. 2d ed. PWS-Kent, 1988.

42. Gilbert, Thomas F., and Marilyn B. Gilbert. *Thinking Metric*. 2d ed. Wiley, 1978.

43. Gnanadesikan, Mirudulla, Richard L. Scheaffer, and Jim Swift. *The Art and Techniques of Simulation*. Seymour, 1986.

44. Goldberg, Kenneth P. *Pushbutton Mathematics: Calculator, Math Problems, Examples, and Activities*. Prentice-Hall, 1982.

45. Goodman, Harvey, et al. *A Guide to Establishing a Science Mathematics Research Program in High School*. New York City Board of Education, 1982.

46. Greenberg, Marvin. *Euclidean and Non-Euclidean Geometries: Development and History*. 2d ed. Freeman, 1980.

47. Gudder, Stanley. *A Mathematical Journey*. McGraw, 1976.

48. Guy, Richard K. *Unsolved Problems in Number Theory*. Springer-Verlag, 1982.

49. Henderson, Linda Dalrymple. *The Fourth Dimension and Non-Euclidean Geometry in Modern Art*. Princeton, 1983.

50. Henry, Boyd. *Every Number Is Special*. Seymour, 1988.

51. Herstein, I. N. *Abstract Algebra*. Macmillan, 1986.

52. Hess, Adrien. L. *Four-Dimensional Geometry—an Introduction*. NCTM, 1977.

53. Hilton, Peter, and Jean Pedersen. *Build Your Own Polyhedra*. Addison-Wesley, 1988.

54. Hinton, Charles H. *Speculations on the Fourth Dimension: Selected Writings of C. H. Hinton*. Dover, 1980.

55. Hobson, E. W. *Squaring the Circle*. Chelsea, 1953.

56. Hua, Lo-Keng, and Yuan Wang. *Applications of Number Theory to Numerical Analysis*. Springer-Verlag, 1981.

57. Huck, Schuyler, and Howard Sandler. *Statistical Illusions: Problems and Solutions*. Harper, 1984.

58. Hunkins, Dalton R., and Larry R. Mugridge. *Applied Finite Mathematics*. 2d ed. Prindle, Weber & Schmidt, 1985.

59. Huntley, H. E. *The Divine Proportion: A Study in Mathematical Beauty*. Dover, 1970.

60. Huntsberger, David V., and Patrick Billingsley. *Elements of Statistical Inference*. 5th ed. Allyn & Bacon, 1981.

61. Jacobs, R. Harold. *Mathematics, a Human Endeavor*. Freeman, 1982.

62. Jaffe, Abram J., and Herbert F. Spirer. *Misused Statistics: Straight Talk for Twisted Numbers*. Dekker, 1987.

63. Jain, M. K., S. R. K. Iyengar, and Rajendra K. Jain. *Numerical Methods for Scientific and Engineering Computation*. Wiley, 1985.

64. Johnson, Lee W., and R. Dean Riess. *Numerical Analysis*. 2d ed. Addison-Wesley, 1982.
65. Kenney, Margaret J., and Stanley J. Bezuszka. *Tessellations Using Logo*. Seymour, 1988.
66. Kirk, Roger E. *Elementary Statistics*. 2d ed. Brooks/Cole, 1984.
67. Kline, Morris. *Mathematics: An Introduction to Its Spirit and Use*. Freeman, 1979.
68. Koblitz, Ann Hibner. *A Convergence of Lives: Sofia Kovalevskaia, Scientist, Writer, Revolutionary*. Birkhäuser, 1983.
69. Kooker, Earl W., and George Paul Robb. *Introduction to Descriptive Statistics*. Thomas, 1982.
70. Kosniowski, Czes. *Fun Mathematics on Your Microcomputer*. Cambridge, 1983.
71. Kostovskii, A. *Geometrical Constructions Using Compass Only*. Seymour, 1986.
72. Kotz, Samuel, and Norman L. Johnson. *Encyclopedia of Statistical Sciences*. Vol. 1-8. Wiley, 1988.
73. Krause, Eugene F. *Taxi-Cab Geometry*. Dover, 1986.
74. Landwehr, James M., and Ann E. Watkins. *Exploring Data*. Seymour, 1986.
75. Landwehr, James M., Jim Swift, and Ann E. Watkins. *Exploring Surveys and Information from Samples*. Seymour, 1986.
76. Lindgren, Bernard W,. and Donald A. Berry. *Elementary Statistics*. Macmillan, 1981.
77. Lowe, U. *Alfred North Whitehead*. Johns Hopkins, 1985.
78. Luckiesch, M. *Visual Illusions*. Seymour, 1986.
79. Lyng, Merwin. *Dancing Curves: A Dynamic Demonstration of Geometric Principles*. NCTM, 1978.
80. Mandelbrot, Benoit B. *The Fractal Geometry of Nature*. Freeman, 1982.
81. Manvel, B. *Introduction to Combinatorial Theory*. Wiley, 1984.
82. Maron, Melvin J. *Numerical Analysis: A Practical Approach*. 2d ed. Macmillan, 1987.
83. Marr, Richard. *Four Dimensional Geometry*. Houghton Mifflin, 1970.
84. Martin, George Edward. *Transformation Geometry: An Introduction to Symmetry*. Springer-Verlag, 1982.
85. *Mira Activities*. Creative Publications.
86. Moise, Edwin. *Elementary Geometry from an Advanced Standpoint*. Addison-Wesley, 1974.
87. Moreau, Rene. *The Computer Comes of Age: The People, The Hardware, and the Software*. MIT, 1984.
88. Morris, John L. *Computational Methods in Elementary Numerical Analysis*. Wiley, 1983.
89. Mott, Joe L., Abraham Kandel, and Theodore P. Baker. *Discrete Mathematics for Computer Scientists and Mathematicians*. 2d ed. Prentice-Hall, 1986.
90. Newman, Claire M., Thomas E. Obremski, and Richard L. Scheaffer. *Exploring Probability*. Seymour, 1986.
91. O'Shea, Tim. *Learning and Teaching with Computers: Artificial Intelligence in Education*. Prentice-Hall, 1983.
92. Oldknow, Adrian, and Derek Smith. *Learning Mathematics with Micros*. Wiley, 1983.
93. Olson, Alton T. *Mathematics through Paper Folding*. NCTM, 1987.
94. Osen, Lynn. *Women in Mathematics*. MIT, 1974.
95. Packel, Edward. *The Mathematics of Games and Gambling*. MAA, 1981.
96. Pearson, Egan, and Maurice Kendall. *Studies in the History of Statistics and Probability*. Hafner, 1970.
97. Perl, Teri. *Mathematical Equals: Biographies of Women Mathematicians and Related Activities*. Addison-Wesley, 1978.
98. Peter, Rozsa. *Playing with Infinity*. Seymour, 1986.
99. Phillips, John L., Jr. *How to Think about Statistics*. Seymour, 1986.
100. Pohl, Victoria. *How to Enrich Geometry Using String Designs*. NCTM, 1986.
101. Posamentier, Alfred S., and William Wernick. *Advanced Geometric Constructions*. Seymour, 1988.
102. Ritchie, David. *The Binary Brain: Artificial Intelligence in the Age of Electronics*. Little, Brown, 1984.

103. Robinson, Enders A. *Statistical Reasoning and Decision Making*. Goose Pond, 1981.

104. Rosen, Kenneth H. *Elementary Number Theory and Its Applications*. Addison-Wesley, 1985.

105. Row, Sundara T. *Geometric Exercises in Paper Folding*. Seymour, 1986.

106. Rucker, Rudy. *Infinity and the Mind: The Science and Philosophy of the Infinite*. Birkhäuser, 1982.

107. Runion, Garth E., and James R. Lockwood. *Deductive Systems: Finite and Non-Euclidian Geometries*. NCTM, 1978.

108. Runyon, Richard P. *Winning with Statistics: A Painless First Look at Numbers, Ratios, Percentages, Means and Inferences*. Addison-Wesley, 1977.

109. Schank, Roger C., and Peter G. Childers. *The Cognitive Computer: On Language, Learning, and Artificial Intelligence*. Addison-Wesley, 1984.

110. Schroeder, M. R. *Number Theory in Science and Communication*. Springer-Verlag, 1984.

111. Schwartz, Judah L., and Michael Yerushalmy. *The Geometric Presupposer*. Sunburst.

112. ———. *The Geometric Supposers: Triangles; Quadrilaterals; Circles*. Sunburst.

113. Selkirk, K. E. *Pattern and Place: An Introduction to the Mathematics of Geography*. Cambridge, 1982.

114. Seymour, Dale, and Reuben Schadler. *Creative Constructions*. Creative Publications, 1974.

115. Seymour, Dale, Linda Silvey, and Joyce Snider. *Line Designs*. Creative Publications, 1974.

116. Seymour, Dale. *Geometric Design*. Seymour, 1988.

117. Shurkin, Joel N. *Engines of the Mind: A History of the Computer*. 1st ed. Norton, 1984.

118. Smart, James R. *Modern Geometries*. 2d ed. Brooks/Cole, 1978.

119. Smith, David E. *History of Mathematics*. 2 vols. Dover or Seymour, 1923.

120. Smith, J. Karl. *The Nature of Mathematics*. 5th ed. Brooks/Cole, 1986.

121. Snover, Stephen and Mark Spikell. *Brain Ticklers: Puzzles and Pastimes for Programmable Calculators and Personal Computers*. Prentice-Hall, 1981.

122. Solow, Daniel. *How to Read and Do Proofs: An Introduction to Mathematical Thought Processes*. Wiley, 1982.

123. ———. *Reading, Writing, and Doing Mathematical Proofs—Proof Techniques for Geometry for Advanced Math*. Wiley, 1982.

124. Somerville, Edith. *A Rhythmic Approach to Mathematics*. 1906. Reprint. Classics in Mathematics Education, vol. 5. NCTM, 1975.

125. Sondheimer, E. S. Rogerson. *Numbers and Infinity: Historical Account of Mathematical Concepts*. Cambridge, 1981.

126. Spencer, David D. *Computers in Number Theory*. Computer Science, 1982.

127. ———. *The Illustrated Computer Dictionary*. Merrill, 1980.

128. Stein, Dorothy. *Ada, a Life and a Legacy*. Cambridge, 1985.

129. Stein, Sherman. *The Man-Made Universe*. Freeman, 1976.

130. Stolyar, Abram Aronovich. *Introduction to Elementary Mathematical Logic*. Dover, 1984.

131. Thomas, David A. *Math Projects for Young Scientists*. Watts, 1988.

132. Thompson, Patrick W. *Motions—a Microworld for Investigating Motion Geometry*. Cosine, 1984.

133. Townsend, M. Stewart. *Mathematics in Sport*. Wiley, 1984.

134. Tucker, Alan. *Applied Combinatorics*. 2d ed. Wiley, 1984.

135. Vandergraft, James S. *Introduction to Numerical Computations*. 2d ed. Academic, 1983.

136. Waits, Bert K., and Franklin Demana. *Using Graphing Calculators to Enhance the Teaching and Learning of Precalculus Mathematics*. Ohio State, 1989.

137. Walker, Helen. *Studies in the History of Statistical Methods*. Ayer, 1975.

138. Wells, David. *The Penguin Dictionary of Curious and Interesting Numbers*. Viking Penguin, 1987.

139. Wenninger, Magnus J. *Polyhedron Models*. Cambridge or Seymour, 1971.

140. ———. *Spherical Models*. Cambridge or Seymour, 1979.

141. Wilder, Raymond L. *Introduction to the Foundations of Mathematics*. 3d ed. Krieger, 1980.

142. Williams, David E. *Mathematics Teacher's Complete Calculator Handbook.* Prentice-Hall, 1984.

143. Williams, Michael R. *A History of Computing Technology.* Prentice-Hall, 1985.

144. Willson, John. *Mosaic and Tessellated Patterns.* Dover, 1983.

145. Winter, John. *String Sculpture.* Creative, 1986.

146. Wylie Jr., C. R. *101 Puzzles in Thought and Logic.* Dover, 1957.

147. Yates, Robert C. *The Trisection Problem.* 1942. Reprint. Classics in Mathematics Education, vol. 3. NCTM, 1971.

148. Zehna, Peter W. *Probability by Calculator: Solving Probability Problems with the Programmable Calculator.* Prentice-Hall, 1982.

149. Zeisel, Hans. *Say It with Figures.* Harper & Row, 1985.

150. NCTM. *Computers in Mathematics Education.* 1984 Yearbook. The Council, 1984.

151. ———. *Curves and Their Properties.* 1943. Reprint. Classics in Mathematics Education, vol. 4. The Council, 1974.

152. ———. *Developing Computational Skills.* 1978 Yearbook. The Council, 1978.

153. ———. *The Growth of Mathematical Ideas, Grades K–12.* Twenty-fourth Yearbook. The Council, 1959.

154. ———. *The Ideas of Algebra, K–12.* 1988 Yearbook. The Council, 1988.

155. ———. *Insights into Modern Mathematics.* Twenty-third Yearbook. The Council, 1957.

156. ———. *Instructional Aids in Mathematics.* Thirty-fourth Yearbook. The Council, 1973.

157. ———. *Learning and Teaching Geometry, K–12.* 1987 Yearbook. The Council, 1987.

158. ———. *A Metric Handbook for Teachers,* edited by Jon L. Higgins. The Council, 1974.

159. ———. *The Metric System of Weights and Measures.* Twentieth Yearbook. The Council, 1948.

160. ———. *Multi-Sensory Aids in the Teaching of Mathematics.* Eighteenth Yearbook. The Council, 1945.

161. ———. *Organizing for Mathematics Instruction.* 1977 Yearbook. The Council, 1977.

162. ———. *A Source Book of Mathematical Applications.* Seventeenth Yearbook. The Council, 1942.

163. ———. *Teaching Statistics and Probability.* The Council, 1981.

164. ———. *Topics for Mathematics Clubs.* 2d ed. The Council, 1983.

Articles

165. Alexander, F. D. "The Metric System, Let's Emphasize Its Use in Mathematics." *Arithmetic Teacher* 20 (May 1973): 395–96.

166. Arcvi, A. "Using Historical Materials in the Math Classroom." *Arithmetic Teacher* 35 (December 1987): 13–16.

167. Bidwell, J. K. "Using Reflections to Find Symmetric and Asymmetric Patterns." *Arithmetic Teacher* 34 (March 1987): 10–15.

168. Bohen, H. "Paper Folding and Equivalent Fractions: Bridging a Gap." *Arithmetic Teacher* 18 (April 1971): 245–49.

169. Brieske, Tom. "Visual Thinking with Translations, Halfturns, and Dilations." *Mathematics Teacher* (September 1984): 466–69.

170. Burk, Frank. "Some Interesting Consequences of a Hyperbolic Inequality." College *Mathematics Journal* 17 (January 1986): 75-76.

171. Good, Robert C. Jr. "The Binary Abacus: A Useful Tool for Explaining Computer Operations." *Journal of Computers in Mathematics and Science Teaching* 5 (Fall 1985): 34–37.

172. Gregory, Richard. "Visual Illusions." *Scientific American* 219 (November 1968): 66–76.

173. Haak, Sheila. "Using the Monochord: A Classroom Demonstration on the Mathematics of Musical Scales." *Mathematics Teacher* 75 (March 1982): 238–244.

174. Hess, Lindsay L. "Models Depicting Groups of Movements." *School Science and Mathematics* 58 (November 1958): 585–92.

175. Hess, Adrien L. "Viewing Diagrams in Four Dimensions." *Mathematics Teacher* 64 (March 1971): 247–48.

176. Hess, Adrien L., and Carl Diekhans. "The Number of Segments Needed to Extend a Cube to *N* Dimensions." *Mathematics Magazine* 43 (September 1970): 187–92.

177. Hoffman, Dale T. "Smart Soap Bubbles Can Do Calculus." *Mathematics Teacher* 72 (May 1979): 377–85, 389.

178. Horak, Virginia M., and Willis J. Horak. "Geometric Proofs of Algebraic Identities." *Mathematics Teacher* 74 (March 1981): 212–16.

179. Kimberling, Clark. "Microcomputer-Assisted Discoveries: Euclidean Algorithm and Continued Fractions." *Mathematics Teacher* 76 (October 1983): 510–12.

180. Kluepfel, Charles. "When Are Logarithms Used?" *Mathematics Teacher* 74 (April 1981): 250–53.

181. Laffan, Anthony J. "Polyhedron Candles: Mathematics and Craft." *Arithmetic Teacher* 28 (November 1980): 18–19.

182. Lund, Charles. "Pascal's Triangle and Computer Art." *Mathematics Teacher* 72 (March 1979): 170–84.

183. Manuel, George, and Amalia Santiago. "An Unexpected Appearance of the Golden Ratio." *College Mathematics Journal* 19 (March 1988): 168–170.

184. Masalski, William J. "Equation Plotter I." *Journal of Computers in Mathematics and Science Teaching* 4 (Spring 1985): 26–30.

185. Niman, John and Robert D. Postman. "Probability on the Geoboard." *Arithmetic Teacher* 20 (March 1973): 167–70.

186. Poggi, Jeanlee M. "An Invitation to Topology." *Arithmetic Teacher* 33 (December 1985): 8–11.

187. Rector, Robert E. "Game Theory: An Application of Probability." *Mathematics Teacher* 8 (February 1987): 138–42.

188. Salisbury, Andrew John. "Some Strategies Games Using Desargue's Theorem." *Mathematics Teacher* 75 (November 1982): 652–53.

189. Schwartzman, Steven. "Factoring Polynomials and Fibonacci." *Mathematics Teacher* 76 (January 1986): 54–56.

190. Slawsky, Norman. "The Artist as Mathematician." *Mathematics Teacher* 70 (April 1977): 298–308.

191. Thompson, Patrick W. "A Piagetian Approach to Transformation Geometry via Microworlds." *Mathematics Teacher* 78 (September 1985): 465–471.

192. Vest, Floyd. "Some Notes on Chase's Method of Subtraction and Application to Base Minus Ten Arithmetic." *ICTM* 7 (Fall 1978): 25–28.

193. Wiener, Joseph. "Bernoulli's Inequality and the Number *e*." *College Mathematics Journal* 16 (November 1985): 399–400.

APPENDIX A

MATHEMATICS PROJECT PROGRESS REPORT*

Student name ______________________________ Date ________________

1. Project title: ____

____

2. Mathematics content involved: ________________________________

____

3. Resources

   For journals: article title, journal name, volume and number, date of publication, pages used.

   For books: title, location of publisher, publisher, copyright date, pages used.

____

____

4. Questions that this project addresses:

(a)____

(b)____

(c)____

(d)____

5. Time-line for completion of the project:

Activity to be completed	Deadline date

6. Attach whichever is appropriate for your project: outline of written report or sketches of visual aids.

7. List any questions or concerns with which you need assistance:

(a)____

(b)____

(c)____

* Based on "Geometry Project—Progress Report" by Rheta Rubenstein, Renaissance High School, Detroit, Michigan.

APPENDIX B

TIPS FOR ORAL PRESENTATION OF PROJECTS

1. Be well prepared and well rehearsed—practice in front of a friendly critic.
2. Dress appropriately, but so that you feel good—the better you feel, the better you'll do.
3. Take time for yourself before you present—5 or 10 minutes.
4. Remove gum, toothpicks, and so on, from your mouth.
5. Nerves and adrenaline are OK, just don't let them dominate your presentation.
6. Use statements on notecards as reminders—avoid reading directly from notecards unless you need a precise quote.
7. Start with a grabber—a strong statement or activity that captures the attention of your audience.
8. Speak slowly and clearly—you will generally speak faster than you perceive.
9. Use a voice level that is adequate for the room; you can't impress the audience if they are not able to hear you.
10. Modulate your voice—if you sound enthusiastic about the project, the listeners will be excited also.
11. Establish eye contact with your audience.
12. Smile occasionally, be animated, use hand gestures—body language often says more than words.
13. Stand to the side of your visuals—"seeing" reinforces your verbal statements.
14. *Above all,* remember that you are the *expert* on your *project*—and know more about your subject than most or all of your audience. Feel good about yourself and all the work you invested in the preparation of the project.

NOTES